FIRST EDITION

Congratulations!

You now have the "tool-kit" to build an Outstanding Consulting Engineering Practice. I wish you much success.

Jack Gaskell

The **"Outline Guide"** to CONSULTING ENGINEERING

How To Start & Manage An Outstanding Consulting Engineering Practice

John D. Gaskell
Retired Professional Engineer

Professional Value Books, Inc.
110 Shore Court, Unit 304, North Palm Beach, FL 33408
TheEngineersResource.com

The "Outline Guide" to CONSULTING ENGINEERING

International Standard Book Numbers (ISBN)
Hardcover: 978-0-9964197-0-3
Softcover: 978-0-9905120-2-8
E-Book: 978-0-9905120-3-5

Interior Designer: JERA Publishing (Stephanie Anderson)
Cover Designer: JERA Publishing (Jason Orr)

Publisher: Professional Value Books, Inc., http://ProfessionalValueBooks.com.
110 Shore Court, Unit 304, North Palm Beach, FL 33408 U.S.A.

Author: John D. Gaskell http://TheEngineersResource.com

Library of Congress Control Number: 2014953977

DEDICATION

This book is dedicated to the memory of John W. King, PE, my mentor, my partner, my friend, and the most honest and generous man that I have ever known.

This book is also dedicated to the memory of my wife, Patricia M. Gaskell (Levesque), the love of my life, the mother of my children, and the smartest person that I have ever known.

They are both missed by many.

Lastly, this book is also dedicated to my grandson(s). Danny Butler was my first grandchild and I have two other "honorary" grandsons, his cousins Billy and Robert. They all call me Pe'Pa and always greet me with a hug. They are delightful and I am truly blessed.

FOREWORD

In his *Outline Guide to Consulting Engineering*, Jack Gaskell provides readers with a dose of wisdom that can't be had by just Googling on the Internet. Jack possesses a rare combination of business acumen and engineering knowledge which made his career a success and, now, his writing of value. In the book, he is offering those just entering the profession the benefits of a lifetime of experience and hard-learned lessons. Jack teaches readers certain essential skills that are timeless—these do not change with technology: how to treat clients in various situations, the art of professional selling, and making intelligent back-office decisions. The book is a must read for the young consultant and a perfect reference guide for experienced engineers.

James R. Fink, PE, Kleinholz Inc., Consulting & Forensic Engineers
kleinholzinc.com

PREFACE

The "Outline Guide" to CONSULTING ENGINEERING is a condensed and abridged summery of the detailed advice provided in the new book: *The "Complete Guide" to CONSULTING ENGINEERING* by John D. Gaskell, Retired Professional Engineer. In essence, readers discover "step-by-step" how to start & manage an "outstanding" Engineering Practice and exactly how to gain a reputation as an expert in their specialty.

This is both a concise handbook for new engineers and a constant reference manual for seasoned professionals.

The "Outline Guide" is the result of requests from many readers to narrow the focus to offer the "highlights" of the recommendations.

In addition, it is anticipated that companies which provide products or services to engineers will present this "Outline Guide" to engineers as a Premium-Gift to promote new business and to thank engineers that already support them.

If purchased in substantial quantities, a Special Edition named in honor of the sponsor could be printed: *(Sponsor's) "Outline Guide" to CONSULTING ENGINEERING.*

John D. Gaskell, Retired Professional Engineer
TheEngineersResource.com

ACKNOWLEDGMENTS

FOR AUTHORIZATION TO WRITE THIS BOOK:
Thomas E. Lent, President and CEO of Thielsch Engineering, Inc., for relieving me of the "non-disclosure" requirements of our purchase and sale agreement when Thielsch Engineering purchased Gaskell Associates, Ltd. I am also grateful to Tom for the professional way that he treated me while I served as one of his Division Presidents, his friendship, and his encouragement in writing this book.

FOR GATHERING INFORMATION:
When I retired from Thielsch Engineering, I had no plans to write a book and did not take copies of anything. But former employees and clients were willing to search through the archives to provide information that was greatly helpful.

Joe Beretta, Bob Bravo, Greg Driggers, Jim Fink, Roger Harris, Gary Hebner, Dana Newbrook, Jackie Rand, and Duncan Speel.

FOR PREPARING THE COVER BACKGROUND:
Evan Oliveira and Francesco Bianco

FOR REVIEWING & EDITING MY CHAPTER ON "FORENSIC ENGINEERING," and for much advice and encouragement over the years as a colleague and friend.

Marvin M. Specter, PE, LS, PP, F.NAFE, F.ITE, F.NSPE, Executive Director Emeritus of the National Academy of Forensic Engineers (NAFE).

FOR REVIEWING MY MANUSCRIPT:
Several people generously gave their time to review and provide constructive advice that greatly improved the final publication of this book.

Joseph R. Beretta, President and CEO, The RGB Corporation; Michelle, Daniel, Danny, and Jillian Butler; Robert C. Bravo, PE, Griffith and Vary, Inc.; Pamela M. Cox; Raymond A. DeCesare, AIA, retired; James R. Fink, PE, Kleinholz, Inc., Consulting & Forensic Engineers; John T. Gaskell, ESQ.; Gary Hebner, Senior Commissioning Agent, Stephen Turner, Inc.; Dana M. Newbrook, AIA, NCARB, a i designs, Ltd.; Matthew J. Viana, PE, Senior Engineer, Millstone Engineering, P.C.; and Robert Steven Wilkinson, PE, Wilkinson Associates, Inc.

ABOUT THE AUTHOR

John D. Gaskell is a retired consulting engineer who wishes to share his 35 years of experience with both young engineers thinking of starting their own practice and those old hands who wish to improve their results.

In essence, Jack's readers discover "step-by-step" how to start and manage an "*outstanding*" engineering practice and exactly how to gain a reputation as an expert in their specialty.

He was President of Gaskell Associates Consulting Engineers, now a Division of Thielsch Engineering, Inc. Jack is a graduate of Wentworth Institute, with an associate's degree in Electrical Engineering Technology, and the University of Rhode Island, with a Bachelor of Science degree in Electrical Engineering.

Jack was a member of the following professional organizations: Rhode Island Society of Professional Engineers, Providence Engineering Society, Electrical League of Rhode Island, Illuminating Engineering Society of North America, National Electromagnetic Field Testing Association, and the National Academy of Forensic Engineers.

He is a past President of the Rhode Island Society of Professional Engineers, past President of the Rhode Island Chapter of the Illuminating Engineering Society, past Director of the Electrical League of Rhode Island, and past Chairman of the Electrical Code Sub-Committee of the Rhode Island Building Code Standards Committee.

Jack was honored by being selected "Engineer of the Year" by the Rhode Island Society of Professional Engineers; "Man of the Year" by the Electrical League of Rhode Island; and "Freeman Award" recipient by the Providence Engineering Society for the purpose of recognizing major achievements in engineering.

He spends winters at his home in North Palm Beach, FL, enjoying the warm weather and losing at the game of pool. Summers are spent at his home in Warwick, RI, enjoying his children, grandchildren, and many great friends.

Jack also has three other books under way. Look for them!

GUARANTEE

The publisher is so convinced that you will be 100% satisfied with this book that we provide a full money back, no questions asked guarantee. Just notify the publisher at Professional Value Books, Inc., 110 Shore Court, Unit 304, North Palm Beach, FL 33408, provided that the book was not customized for you and was purchased from the website: TheEngineersResource.com.

DISCLAIMER

The author, John D. Gaskell, is a retired/inactive/no longer licensed engineer. He feels qualified to tell you how he started and managed his consulting engineering practice and to give general advice. But he is not an attorney, accountant, insurance expert, or any other kind of business professional and can only give general guidance, which should always be confirmed by an attorney or other appropriate professionals before proceeding. He has expended considerable effort in the preparation of his books, articles, and blogs. He believes, to the best of his knowledge, that the information contained therein is accurate. Nothing should be construed as legal advice, and he offers his apologies, in advance, for the inevitable blunders.

CONTENTS

PART 1
PREPARATION

CHAPTER 1

BECOMING A PROFESSIONAL ENGINEER

INTRODUCTION

Learn about the exciting field of consulting engineering and how to determine if this is the right career choice for you. Discover step-by-step how to become a licensed professional engineer, including educational requirements, experience, and examinations. Find out the benefits and drawbacks of consulting engineering as well as the advantages and disadvantages of owning your own practice.

An engineer is "a person who has scientific training and who designs and builds complicated products, machines, systems, or structures".

A Professional Engineer (PE in the U.S.) is one who has attained a credential that permits him to provide engineering services to the general public.

Consulting engineers are individuals who, because of training in one or more engineering specialties, are licensed professional engineers in private practice. They serve private and public clients in ways ranging from brief consultations to complete design and coordination of projects. They are often the technical liaison between architects, process specialists, contractors, suppliers and the client. A consulting engineer can provide general consultation, feasibility reports, design, cost estimates, rate studies, project development, patent assistance, and preparation of environmental impact statements.

After I passed the PE exam, I was registered as a professional engineer, which meant that I could certify (attach my PE seal/stamp) to any type of engineering documents, as long as I was willing to show competence in that area of engineering. In recent years, engineers are registered in one or more standard engineering specialties and their practice is limited to those areas of work.

OUTLINE SUMMARY

- **Consider engineering:** Those who have good science and mathematical grades and interests should consider engineering.
- **Shadow:** Spend a day with an engineer of the discipline that interests you.
- **College selection:** Make sure that the college of your choice is ABET-EAC accredited.
- **FE exam:** In your senior year of college, take the FE written examination in the fundamentals of engineering.
- **Gain experience:** A minimum of four (4) years of "certifiable" experience in engineering work is required to be eligible to take the PE exam.
- **PE exam:** Take the PE exam as soon as you are eligible. Not everyone passes the first time. Don't give up.
- **Know the laws:** Comply with the Professional Registration Laws in all the states where you will practice.
- **Consider consulting engineering:** Carefully ponder the pros and cons of being a "consulting engineer":
  - Benefits: Full involvement with all aspects of interesting projects while not being tied to a desk.
  - Drawbacks: Being in an adversarial position in a cyclical industry, the stress of deadlines, the pressure to be profitable, and the dilemma of either too much or too little work.
- **Consider ownership:** Carefully mull over the pros and cons of "owning" a consulting engineering practice:
  - Benefits: Making all of the final decisions; never being laid-off; keeping all of the profits; and being able to sell a valuable asset upon retirement.
  - Drawbacks: Personal and family sacrifices; difficulty of dealing with employees and acceptance of any losses.

CHAPTER 2

PREPARING FOR YOUR OWN FIRM

INTRODUCTION

Don't just "stumble" into becoming a consulting engineer. Plan for it. Learn how to systematically search for and get the right job in a consulting engineering firm, one that will qualify you to take the PE exam and give you the knowledge and experience to eventually head-up your own firm. Detect the secrets of being properly prepared for your job interview. Discover how to become an "exceptional" employee, and see how that will help you when you start your own "outstanding" engineering practice.

Find out how easy it is to transcend from the status of a "nobody" to being "widely known and respected." Uncover the mysteries of getting elected "unopposed" to Engineering Society Boards and being nominated for prestigious awards. With determination and effort, you can establish credentials that will eventually distinguish you from your competitors. The important lesson here is: "It doesn't just happen—you make it happen." Read how!

Clients hire engineers to draw system plans and write specifications for their buildings, including (but not limited to) electrical, mechanical (heating, ventilating, air conditioning, and plumbing), fire protection (sprinklers), structural, and civil. Also, sometimes very narrow specialties are required, such as acoustical engineering. Electrical systems, for example, include: utilities serving the building (power, telephone, data, cable TV, and fire alarm), lighting, power distribution, fire alarm systems, telephone distribution, cable TV distribution, and any other electrical system that the particular building might require.

The consulting engineer is responsible for designing within the client's budget limitations and coordinating with the requirements of utility companies and inspection authorities. At the end of the design phase, the consultant prepares a specification document detailing the material requirements and system functions for his specialty. During bidding, he attends pre-bid meetings, clarifies issues, and prepares addenda for the architect to issue to inform bidders of changes in the requirements. After a contract is awarded, the consultant reviews/approves shop drawings detailing all equipment that the contractor proposes. During the construction phase, he visits the job site to record progress and clarifies the contract documents. At the completion of the construction phase, he prepares a "punch-list" detailing corrections to the work, if needed.

OUTLINE SUMMARY

- **Plan ahead:** Don't "stumble" into becoming a consulting engineer. Plan for it.
- **Get a job:** Work at a consulting engineering practice that will qualify you to take the PE exam and give you the knowledge and experience to head-up your own firm.
- **The job search:** First, prepare a "resume" stressing CAD and BIM, if you can. Make a list of consulting engineering firms in your area and deliver your resume in person. If you don't quickly get a job, stop back to see the same people. Employment needs can change quickly.
- **The interview:** Be prepared by researching the company's website. Make lists including: five skills and qualifications of yours, likely questions that you might be asked and enquires about the company. Ask if you can meet someone in a similar position and the person who will be your immediate supervisor. Send the interviewer a "Thank You" note or e-mail.
- **Start the job right:** Once you have a job, do your best to be "exceptional." Arrive early and leave late. Start a notebook to record information, such as formulas, code rules, contacts, and anything that you might need again.
- **White papers:** A white paper is a report or guide to help readers to understand an issue. When your work involves a new technical issue, read about it and take detailed notes in narrative form, including definitions of the various new terms. Add this "white paper" to your notebook.
- **Save:** Live within your means so that you will have the assets and good credit to be your own boss.
- **Banking:** Establish a relationship with a financial institution.
- **Gain credentials:** Get elected "unopposed" to Engineering Society Boards, and get nominated for prestigious awards that will distinguish you from your competitors.
- **It doesn't just happen:** "You make it happen."

PART 2

PLANNING

CHAPTER 3

START-UP OPTIONS & SPACE PLANNING

INTRODUCTION

Don't procrastinate to the point where your opportunity for your own engineering practice passes you by. Comprehend the pros and cons of each start-up option and opportunity. Find out how to evaluate offering a single discipline or multi-disciplines. Understand how you might benefit from partners and the associated pitfalls. Discover the step-by-step procedures on how to be ready when opportunity knocks.

Learn how to determine how much space your new firm will initially require and how to plan for the future growth of your firm. Gain knowledge of options, including your house, a suburban office, a downtown office, factory/industrial space, and cooperative/shared space. Discover why actually looking at spaces helps to make the right choice. Find out how easy it is to inventory and price the furniture and equipment that your office will require.

For the purpose of an example, let's assume you're starting a new firm with two other partners. You will be a multi-discipline firm offering electrical, HVAC, plumbing, and fire protection (MEPFP). In addition, you will have one full time receptionist (who will also serve as the secretary and bookkeeper) and one other part-time design/CAD person. Also, let's assume that the part-timer will eventually become full-time and one more designer/CAD person will later be added.

Next, take some sketch paper with a grid of ¼ inch = 1'-0" and rough out a sketch meeting your needs. Don't worry about walls or which space needs to be where. Just fit in everything roughly to scale. Figures **3-1** and **3-2** are the results of my effort. My area requirements measure 2,294 sq. ft. Therefore, for this example, 2,000 to 3,000 square feet of space is tentatively needed. However, keep in mind that most start-up firms need to be very cautious about adding full-time employees. So, extra space for moon-lighters (part-timers working after their day jobs) is an important consideration, and it may take two or three part-timers to offset the work of one full-time employee.

SPACE PLANNING (1 OF 2)

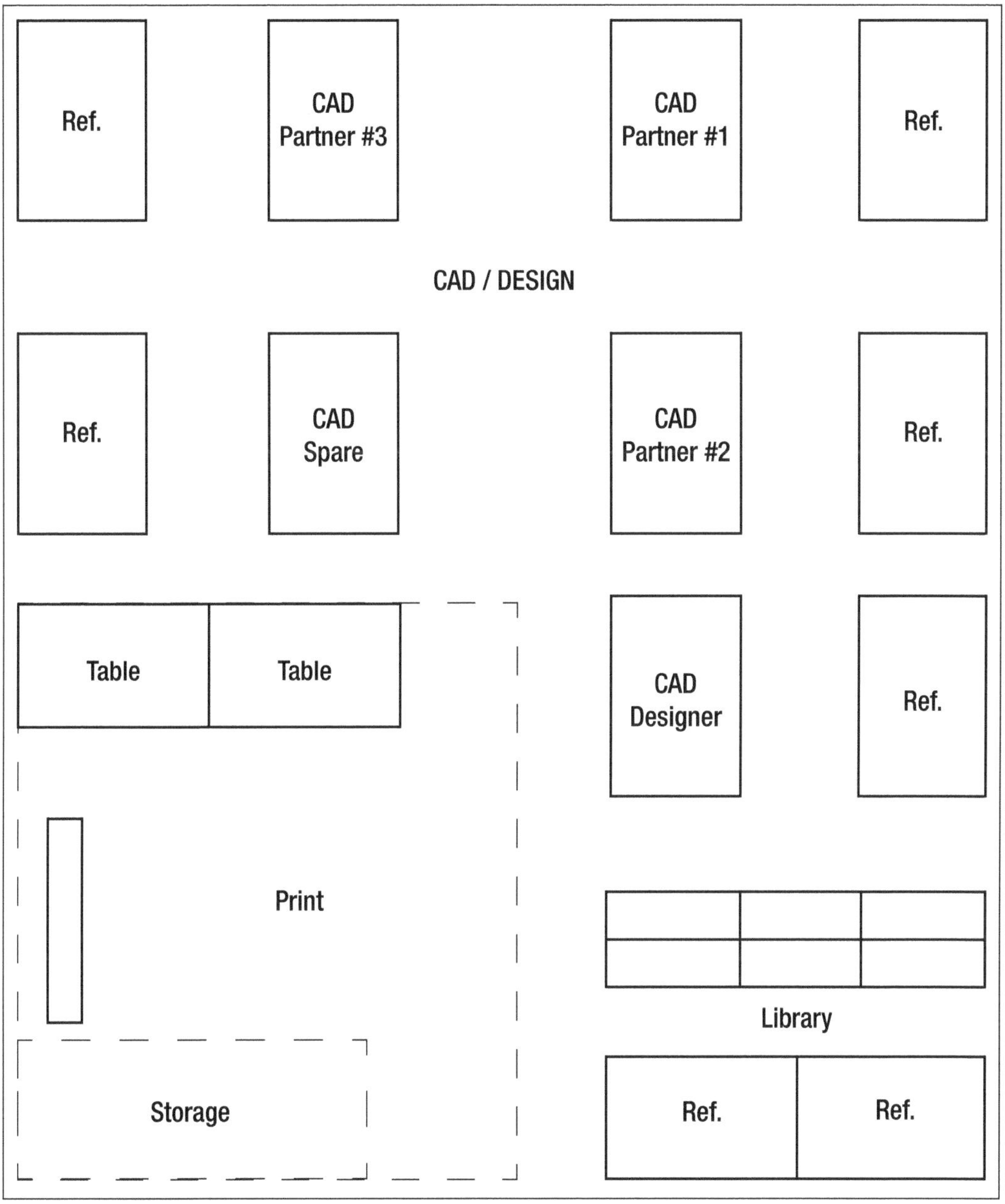

FIGURE 3-1

SPACE PLANNING (2 OF 2)

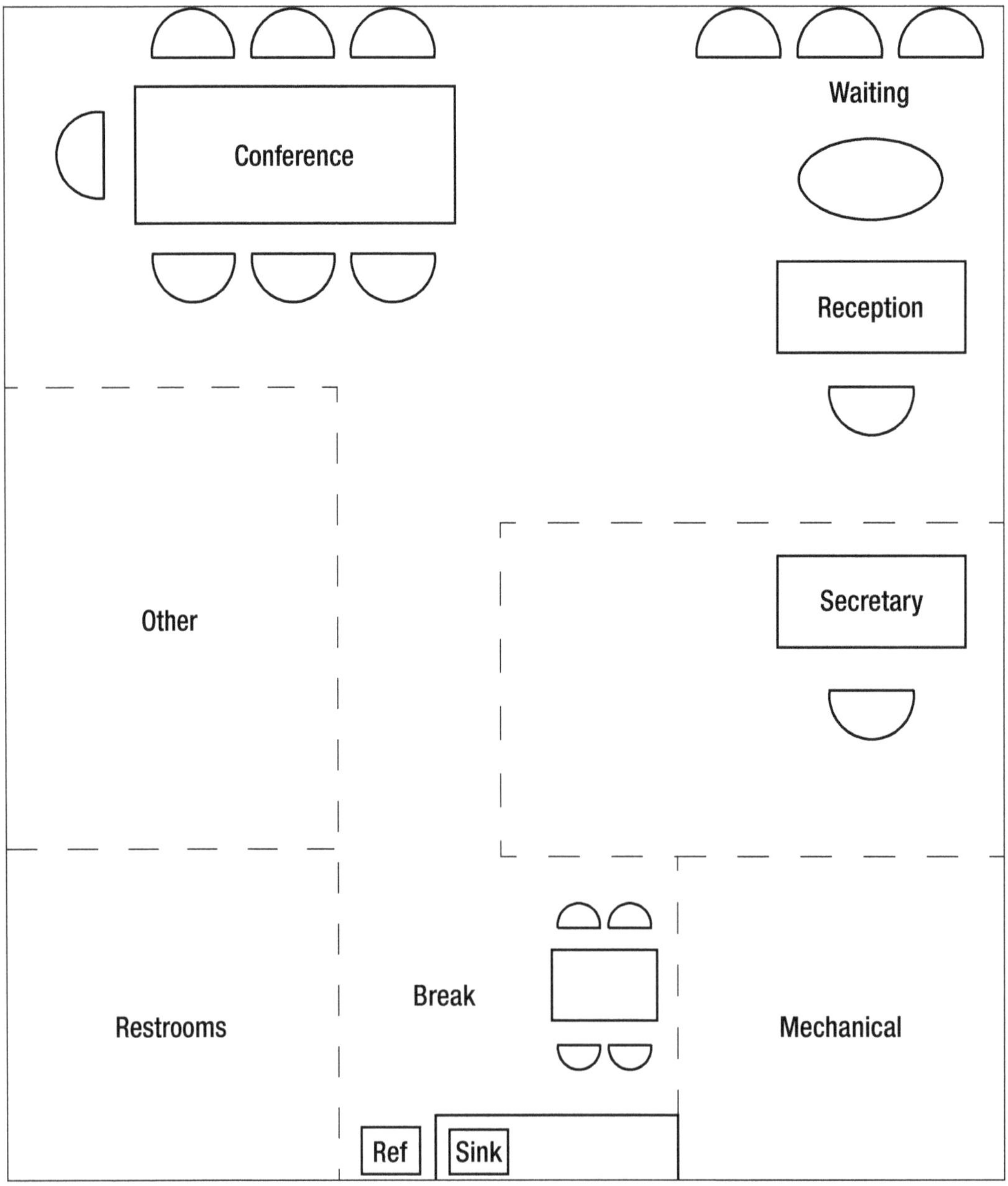

FIGURE 3-2

OUTLINE SUMMARY

Options

- **Cast off:** Within five years of becoming a registered professional engineer, start your own practice.
- **Ethics:** Carefully consider the ethical requirements of your State Boards of Registration when planning, starting, and operating your practice.
- **Start-up options:**
  - Ideally, open your practice at the time that another engineer retires.
  - Offer to buy a practice from a retiring engineer.
  - Become a partner in an existing firm.
  - Quit your job, hang out your "shingle," and start knocking on doors.
- **Sole proprietorship:** Starting as a "Sole Proprietor" is a simple option and is taxed on a personal basis. You get to make all the decisions without the need to get a partner's approval. The disadvantage is that you need to come up with all the funding needs and didn't have a partner to advise against bad ideas.
- **Partnerships:** Carefully consider the pros and cons of being on your own or having partners.
  - Pros: Partners share the workload, help with funding, share the losses, and make a multi-discipline firm more feasible.
  - Cons: Partners share in the profits and the decisions while having a different work ethic, energy levels, and commitment than you do.
- **Disciplines:** Choose which engineering specialties that your firm will offer. Common practice in your geographical area may be your deciding factor. I have observed the following trends. The really big engineering firms offer all disciplines. The multi-discipline firms tend to be heating, ventilating, air conditioning, electrical, plumbing, and fire protection (MEPFP). The single-discipline firms are usually structural, civil, electrical, or mechanical only.

Space Planning

- **Prepare a sketch:** It is not important where rooms are with respect to each other as long as each area meets your needs and is approximately to scale.

- **Office space options:** Consider various rental options and associated costs.
  - Your house: This is your cheapest option and is worth considering if there are no more than three of you and your budget will not support rent. But this doesn't give you a "business" presence or a suitable space for a conference or client meeting.
  - Suburban office: This option avoids the congestion and parking problems of a downtown office and usually has much lower rental rates.
  - Downtown office: If most of your clients are centered in a big city, you may need a downtown office. But you will need to budget "parking costs" for you and your employees.
  - Factory/Industrial: This can be an inexpensive option and may be worth considering but may have drawbacks: maintenance of public spaces, heating systems/AC/Plumbing and electric, neighbors that are noisy or living in their spaces.
  - Cooperative/Shared: A small private space and a shared receptionist/secretary, reception area, conference rooms, break rooms, rest rooms, and copy/fax rooms. As your personal space needs grow, this option soon becomes cost prohibitive.

- **Inventory**: Make a list of furniture and equipment from your sketch, and shop to get actual prices for your Business Plan and to be sure that everything will fit.

- **Used & leasing:** Consider options of used and leasing to conserve cash.

CHAPTER 4

FINANCIAL FORECASTING

INTRODUCTION

You will discover how easy it is to estimate both your Start-Up Costs and First Year Cash Outflows and why you should update your estimate monthly and annually. Learn how to conserve cash by financing or leasing. Find out the secrets of successful firms, like including reimbursable expenses in your contracts. Comprehend why buying-in-bulk is not always the best option.

You will notice how identifying potential clients may be easier than you might think. Realize which people that you already know may be willing to help you. Observe how questions about upcoming projects and expressions of interest can commit these projects to you before the competition ever hears about them. Come to know how much proposed projects will generate in fees for you and when these fees will likely be received.

Example – (Start-Up Costs and First Year Cash Outflows)

In this example, let's estimate the start-up costs and first year cash outflows for the small firm described in Chapter 3 – Start-Up Options & Space Planning, including the furniture and equipment needs depicted in Figures **3-1** and **3-2.**

First, I created the form shown in Figure **4-1** to break down the "start-up" costs. Next, I completed this form by including the costs for my example firm, thus creating Figure **4-2**. (Note that you can download all of these forms from my website at no charge.)

Then, I created the form shown in Figure **4-3** from the categories listed above to include the totals of each category of expense for each month, the total of all expenses for each month, the total of each category of expense for the year, and the grand total of all expenses for the year.

Lastly, I completed this form by including both the start-up costs and the operating cash outflows for my example firm, thus creating Figure **4-4.**

Modify these forms to suit your proposed firm. Resist the temptation to proportion these costs/outflows to your situation; include current pricing applicable to your geographical area. Once in operation, update them monthly and compare changes to your original estimate. This will help to keep your budgeting current and to avoid surprises. Continue to do so each year.

START-UP COSTS

#	EXPENSE ITEMS	EACH	ITEM	SUB TOTALS
Office Fit-Out				
LS	Paint			
LS	Carpet			
LS	Window Treatments			
LS	Break Room Preparation			
	Sub Total (Fit-Out)			
Office Furniture				
4	CAD Furniture			
8	Reference Tables			
12	Book Cases			
16	Chairs			
1	Table (Conference Room)			
1	Table (Waiting Room)			
1	Table (Break Room)			
2	Desks			
8	File Cabinets			
	Sub Total (Office Furniture)			
Equipment				
1	Blueprint Machine			
1	Refrigerator			
1	Copy/Fax machine			
4	CAD Computers and Software			
2	Laptop Computers and Software			
5	Printers			
1	Power Point Projector			
	Sub Total (Equipment)			
Office Phone System				
LS	Equipment			
LS	Instillation			
	Sub Total Phone System			
Other				
	Sub Total Other			
TOTAL START-UP EXPENSES				

FIGURE 4-1

START-UP COSTS EXAMPLE

#	EXPENSE ITEMS	EACH	ITEM	SUB TOTALS
Office Fit-Out				
LS	Paint		1,500	
LS	Carpet		5,000	
LS	Window Treatments		1,500	
LS	Break Room Preparation		2,000	
	Sub Total (Fit-Out)			10,000
Office Furniture				
4	CAD Furniture	300	1,200	
8	Reference Tables	100	800	
12	Book Cases	70	840	
16	Chairs	150	2,400	
1	Table (Conference Room)	300	300	
1	Table (Waiting Room)	200	200	
1	Table (Break Room)	150	150	
2	Desks	300	600	
8	File Cabinets	220	1,760	
	Sub Total (Office Furniture)			8,250
Equipment				
1	Blueprint Machine	4,000	4,000	
1	Refrigerator	200	200	
1	Copy/Fax machine	500	500	
4	CAD Computers and Software	3,150	12,600	
2	Laptop Computers and Software	1,250	2,500	
5	Printers	200	1,000	
1	Power Point Projector	700	700	
	Sub Total (Equipment)			21,500
Office Phone System				
LS	Equipment		1,500	
LS	Instillation		1,000	
	Sub Total Phone System			2,500
Other				
	Sub Total Other			0
TOTAL START-UP EXPENSES				$42,250

FIGURE 4-2

FIRST YEAR CASH OUTFLOWS

EXPENSE	ANNUAL	JAN.	FEB.	MARCH	APRIL	MAY	JUNE	JULY	AUG.	SEPT.	OCT.	NOV.	DEC.
Office Fit-out													
Office Furniture													
Office Equipment													
Software & Up-dates													
Office Phone System													
Professional Salaries													
Clerical Salaries													
Officers Salaries													
Payroll Taxes													
Payroll Services													
Auto													
Travel & Entertainment													
Bluprint Expense													
Office Supplies													
Postage & Delivery													
Dues & Subscriptions													
Professional Development													
General Insurance													
Professional Liability Ins.													
Health Insurance													
Rent													
Utilities													
Bld. Repairs & Maint.													
Equip. Repairs & Maint.													
Telephone & Internet													
Employment Ads.													
Loan Re-payments													
Professional Expenses													
Office Cleaning													
Miscellaneous													
TOTALS													

FIGURE 4-3

FIRST YEAR CASH OUTFLOWS EXAMPLE

EXPENSE	ANNUAL	JAN.	FEB.	MARCH	APRIL	MAY	JUNE	JULY	AUG.	SEPT.	OCT.	NOV.	DEC.
Office Fit-out	$10,000	$10,000											
Office Furniture	$8,250	$8,250											
Office Equipment	$18,000	$18,000											
Software & Up-dates	$3,500	$3,500											
Office Phone System	$2,500	$2,500											
Professional Salaries	$104,000	$4,333	$4,333	$4,333	$4,333	$4,333	$4,333	$13,000	$13,000	$13,000	$13,000	$13,000	$13,002
Clerical Salaries	$52,000	$4,333	$4,333	$4,333	$4,333	$4,333	$4,333	$4,333	$4,333	$4,333	$4,333	$4,333	$4,337
Officers Salaries	$240,000	$20,000	$20,000	$20,000	$20,000	$20,000	$20,000	$20,000	$20,000	$20,000	$20,000	$20,000	$20,000
Payroll Taxes	$47,520	$3,440	$3,440	$3,440	$3,440	$3,440	$3,440	$4,480	$4,480	$4,480	$4,480	$4,480	$4,481
Payroll Services	$2,400	$200	$200	$200	$200	$200	$200	$200	$200	$200	$200	$200	$200
Auto	$5,000	$416	$416	$416	$416	$416	$416	$416	$416	$416	$416	$416	$424
Travel & Entertainment	$5,000	$416	$416	$416	$416	$416	$416	$416	$416	$416	$416	$416	$424
Bluprint Expense	$2,500	$208	$208	$208	$208	$208	$208	$208	$208	$208	$208	$208	$212
Office Supplies	$11,000	$5,500			$1,833				$1,833			$1,834	
Postage & Delivery	$3,000	$250	$250	$250	$250	$250	$250	$250	$250	$250	$250	$250	$250
Dues & Subscriptions	$2,000	$166	$166	$166	$166	$166	$166	$166	$166	$166	$166	$166	$174
Professional Development	$2,000	$166	$166	$166	$166	$166	$166	$166	$166	$166	$166	$166	$174
General Insurance	$3,000	$250	$250	$250	$250	$250	$250	$250	$250	$250	$250	$250	$250
Professional Liability Ins.	$10,000	$833	$833	$833	$833	$833	$833	$833	$833	$833	$833	$833	$837
Health Insurance	$19,000	$1,404	$1,404	$1,404	$1,404	$1,404	$1,404	$1,762	$1,762	$1,762	$1,762	$1,762	$1,766
Rent	$31,408	$4,832	$2,416	$2,416	$2,416	$2,416	$2,416	$2,416	$2,416	$2,416	$2,416	$2,416	$2,416
Utilities	$4,000	$333	$333	$333	$333	$333	$333	$333	$333	$333	$333	$333	$337
Bld. Repairs & Maint.	$2,500	$208	$208	$208	$208	$208	$208	$208	$208	$208	$208	$208	$212
Equip. Repairs & Maint.	$2,000	$166	$166	$166	$166	$166	$166	$166	$166	$166	$166	$166	$174
Telephone & Internet	$10,000	$833	$833	$833	$833	$833	$833	$833	$833	$833	$833	$833	$837
Employment Ads.	$500						$250	$250					
Loan Re-payments	$48,000	$4,000	$4,000	$4,000	$4,000	$4,000	$4,000	$4,000	$4,000	$4,000	$4,000	$4,000	$4,000
Professional Expenses	$6,000	$500	$500	$500	$500	$500	$500	$500	$500	$500	$500	$500	$500
Office Cleaning	$3,800	$300	$300	$300	$300	$300	$500	$300	$300	$300	$300	$300	$300
Miscellaneous	$20,000	$2,500	$1,590	$1,590	$1,590	$1,590	$1,590	$1,590	$1,590	$1,590	$1,590	$1,590	$1,600
TOTALS	$678,878	$97,837	$46,761	$46,761	$48,594	$46,761	$47,211	$57,076	$58,659	$56,826	$56,826	$58,660	$56,907

FIGURE 4-4

Example – (First year cash inflow projection for fees to be received.)

Continuing the example (per Chapter 3), I filled out the cash inflow information in (blank) Figure **4-5**, thus creating Figure **4-6**. The result is a "hypothetical" first year cash inflow projection for fees to be received. Allow 2–4 months after billing to actually receive this income. Again, resist the temptation to proportion these estimates to your situation.

FIRST YEAR CASH INFLOWS

CLIENT PROJECT	TOT. REC.	JAN.	FEB.	MARCH	APRIL	MAY	JUNE	JULY	AUG.	SEPT.	OCT.	NOV.	DEC.
A													
B													
C													
D													
E													
F													
G													
TOTALS													

FIGURE 4-5

FIRST YEAR CASH INFLOWS EXAMPLE

CLIENT	PROJECT	TOT. REC.	JAN.	FEB.	MARCH	APRIL	MAY	JUNE	JULY	AUG.	SEPT.	OCT.	NOV.	DEC.
A														
	Office Building	32,000			3,000	3,000	3,000	3,000	3,000	3,000	3,000	3,000	3,000	5,000
	Elem. School	72,000				8,000	8,000	8,000	8,000	8,000	8,000	8,000	8,000	8,000
	Generator Study	12,000						3,000	9,000					
B														
	Baseball Field	20,000				4,000	5,000	5,000		3,000		3,000		
	Riverfront Park	15,000						4,000	2,000	2,000	2,000	2,000	2,000	1,000
	Unknown	25,000					5,000		5,000	5,000	5,000	5,000		
C														
	High School Stdy.	15,000					4,000	10,000	1,000					
	Unknown	60,000							10,000	10,000	10,000	10,000	10,000	10,000
D														
	Unknown	85,000							10,000	15,000	15,000	15,000	15,000	15,000
	Unknown	20,000									20,000			
	Unknown	25,000								15,000				10,000
E														
	Library	23,000									11,000	2,000		10,000
	Dormitory	24,000						9,000				5,000	5,000	5,000
	Unknown	30,000							10,000			10,000	10,000	
	Unknown	25,000										10,000	5,000	10,000
F														
	Hotel Addition	19,500			7,000							5,000	2,500	5,000
	Motel	32,000					8,000	8,000	8,000	8,000				
	Unknown	10,000											10,000	
	Retail Store	6,000											3,000	3,000
G														
	Fire Alarm	4,500											1,500	3,000
	Ginerator	5,000											5,000	
TOTALS		560,000	0	0	10,000	15,000	33,000	50,000	66,000	69,000	74,000	78,000	80,000	85,000

FIGURE 4-6

OUTLINE SUMMARY

- **Get started**: Use the list of typical costs/outflows. Add/delete items as required to suit your practice.
- **Use two categories:** Divide into "start-up" (onetime) costs and "operating" (monthly) outflows.
- **Conserve cash:** Consider "financing" of furniture and equipment and making monthly payments for things like insurance.
- **Reimbursable:** Don't forget to bill for "reimbursable" expenses like printing, if possible.
- **Limit quantities:** Don't stock large quantities of office supplies.
- **Create & update:** Create a spreadsheet for annual and monthly cash outflows. Update it monthly, and compare changes to your original estimate. Continue to do this each year.
- **Potential clients:** Meet with each likely, potential client, and try to gauge their level of commitment to you for future work.
- **Referrals:** Seek referrals from your contact list, consultants of a different specialty, land surveyors, and contractors in your specialty.
- **Other potential clients:** Review newspapers and search on the internet.
- **Confidentiality:** Unless you have already given notice at your present place of employment, inquiries should be discreet.
- **Make projections:** Ask about projects expected within the next year and their anticipated schedule. Also, try to realistically estimate others.
- **Estimate probable fees:** (See Chapter 12 – Fee Proposals)

CHAPTER 5

FINANCIAL ANALYSIS

INTRODUCTION

Discover why your Cash Flow Analysis and Balance Sheets are your most important financial barometers. Learn how to prepare them and when to update them.

You won't be able to borrow without a cash down payment; understand what is expected. Find out which banking relationships to pursue and who has the greater "approval authority." Become aware of "due diligence" and the role that the SBA plays in the loan process. Be prepared for your meeting with the loan officer, and understand the meanings of security, guarantor, endorser, co-signer, demand note, and receivables financing.

CASH FLOW ANALYSIS

The Cash Flow Analysis that I am proposing in this chapter is somewhat of an informal way of estimating your "cash on hand" at the end of each month.

It ignores the issues like depreciation, capital expense determination, etc. Your CPA will take care of these things when he does your year-end accounting. You need to do a Cash Flow Analysis for your initial Business Plan, your subsequent yearly Business Plan, and for each month that you are in business.

Prepare a final spreadsheet combining the Cash Inflows and Cash Outflows into a projected cash flow. See Figure **5-1** Cash Flow Analysis

Example – Cash Flow Analysis

Continuing the example (per Chapters 3 and 4), I filled in both the Cash Inflow and Start-Up Cost information in (blank) Figure **5-1**, thus creating Figure **5-2**. The result is a "hypothetical" first year cash flow projection. Again, resist the temptation to proportion these estimates to your situation.

CASH FLOW ANALYSIS

ITEM	JAN.	FEB.	MAR.	APR.	MAY.	JUN.	JLY.	AUG.	SEP.	OCT.	NOV.	DEC.
Capital Contribution												
Bank Loan												
Add-Cash inflows												
Less-Cash outflows												
Cash on hand												

FIGURE 5-1

CASH FLOW ANALYSIS EXAMPLE

ITEM	JAN.	FEB.	MAR.	APR.	MAY.	JUN.	JLY.	AUG.	SEP.	OCT.	NOV.	DEC.
Capital Contribution	100,000											
Bank Loan	200,000											
Add-Cash inflows	0	0	10,000	15,000	33,000	50,000	66,000	69,000	74,000	78,000	80,000	85,000
Less-Cash outflows	97,837	46,761	46,761	48,594	46,761	47,211	57,076	58,659	56,826	56,826	58,660	56,907
Cash on hand	202,163	155,402	118,641	85,047	71,286	74,075	82,999	93,340	110,514	131,688	153,028	181,121

FIGURE 5-2

BALANCE SHEETS

The final thing that you need is a "balance sheet" for the start of your practice (opening) and for the end of your first year (closing). Hopefully, "closing" means the closing of your books for your first successful year, not the closing of the doors. See Figure **5-3** Balance Sheet.

Example – Balance Sheets

I combined the results of **4-4**, **4-6**, and **5-2** into blank form **5-3** to create **5-4** Balance Sheet – Example.

BALANCE SHEET

ASSETS	OPENING - January 1st		CLOSING - December 31st	
Current Assets				
Cash (Partner Contributions)				
Cash (Bank Loan)				
Cash on hand				
Accounts Receivable				
Prepaid Expenses				
Total Current Assets				
Fixed Assets				
Furniture & Equipment				
Leasehold improvements				
Less - Accumulated Depreciation				
Total Fixed Assets				
TOTAL ASSETS				
LIABILITIES				
Current Liabilities				
Bank Loan Payable (1 year)				
Accounts Payable				
Accrued Expenses				
Total Current Liabilities				
Long-term Debt				
Bank Loan Payable (beyond 1 year)				
TOTAL LIABILITIES				
CAPITAL				
Partners Loans (Owners starting equity)				
Net income (loss)				
Total Capital				
TOTAL LIABILITIES & CAPITAL				

FIGURE 5-3

BALANCE SHEET EXAMPLE

ASSETS	OPENING - January 1st		CLOSING - December 31st	
Current Assets				
Cash (Partner Contributions)	100,000		0	
Cash (Bank Loan)	200,000		0	
Cash on hand	0		137,042	
Accounts Receivable	0		113,000	
Prepaid Expenses	0		2,374	
Total Current Assets		300,000		252,416
Fixed Assets				
Furniture & Equipment	0		32,250	
Leasehold improvements	0		10,000	
Less - Accumulated Depreciation	0		(4,600)	
Total Fixed Assets		0		37,650
TOTAL ASSETS		300,000		290,066
LIABILITIES				
Current Liabilities				
Bank Loan Payable (1 year)	32,000		37,000	
Accounts Payable	0		25,000	
Accrued Expenses	0		5,000	
Total Current Liabilities		32,000		67,000
Long-term Debt				
Bank Loan Payable (beyond 1 year)		168,000		131,000
TOTAL LIABILITIES		200,000		198,000
CAPITAL				
Partners Loans (Owners starting equity)	100,000		100,000	
Net income (loss)	0		(7,934)	
Total Capital		100,000		92,066
TOTAL LIABILITIES & CAPITAL		300,000		290,066

FIGURE 5-4

INITIAL FUNDING NEEDS

Upon completion of your "Cash Flow Analysis," you have a pretty good guess as to your first year's funding needs. If you have followed my recommendations, this should be a realistic projection. But you need to add a safety factor. Assume that you will need more funding than projected, 20–30% more. I recommend that you include this as an identified item in your Business Plan. It shows that you are being conservative and recognize that "Cash Flow Analysis" is not an exacting science.

TYPES OF FUNDING

There are two types of funding: equity and debt.

Equity

This is what you and your partners have saved and are willing to invest in the business. Lenders will expect at least 30% of your first year funding needs to be "equity" funded.

Debt

This is borrowed money that must be paid back. Normally, the principal and interest are paid back in fixed monthly payments. Be sure that you have estimated these payments in your "Cash Flow Analysis."

LOAN SECURITY

Lenders will not give you a loan unless they are convinced that your business will be successful and there is a strong likelihood that the loan will be repaid. Nevertheless, they want to be covered in case of default.

Often "a personal guarantee commitment" is all that is required. Don't offer more than is necessary. If they ask for more, enquire: "Is that really required"? It may be worth your while to shop around to other financial institutions regarding this issue.

Guarantor

For new businesses, lenders will usually expect you and all of your partners to sign a personal guarantee commitment, even if you are a corporation. This makes you each jointly and severable liable for the loan. (Meaning, you are each "individually" liable for the "full" amount). Try to negotiate a limited guarantee to cover the shortfall, if other securities have been pledged.

Endorser

This makes someone else contingently liable for the loan if you default.

Co-Signer

A co-signer is someone who agrees to be jointly liable for the loan.

Promissory Note

A promissory note is usual and includes the principal amount, interest rate, terms of payment and the maturity date.

Demand Note

A demand note is a written promise to pay the outstanding balance upon demand. Avoid this kind of loan. Suggest a "promissory note" instead.

Mortgage

A lender may require a mortgage against your property as collateral; real estate; car; boat; etc.

Pledge of Stocks or Bonds

The lender will usually consider 50 to 75% of the value as security and additional security may be required if the value of these assets declines.

OUTLINE SUMMARY

- **Cash flow analysis**: Estimate your "cash on hand" at the end of each month by preparing a cash flow analysis. You need it for your initial Business Plan, your subsequent yearly Business Plan, and for each month that you are in business.
- **Balance sheet:** The final thing that you need for your Business Plan is a "balance sheet" for the start of your practice (opening) and for the end of your first year (closing). Your balance sheet is your projection of how you expect your firm to grow financially for your first year and will be your key to financing. Hopefully, when you redo it using actual results, you will have exceeded expectations.
- **Buffer:** Assume that you will need more funding than projected, 20–30% more. I recommend that you include this as an identified item in your Business Plan. It shows that you are being conservative and recognize that "Cash Flow Analysis" is not an exacting science.
- **Equity funding:** Lenders will expect at least 30% of your first year funding needs to be "equity" funded.
  - Gifts: Don't overlook the obvious: parents, in-laws, and close relatives.
  - Savings: Passbook savings accounts, CDs, stocks and bonds.
  - Second Mortgage: Long-term, low interest, financing with tax deductible interest.
  - Credit Card: Possible but not recommended.
  - Venture Capitalists, Silent Partners & Angel Investors: Unlikely with pitfalls.
  - Crowdfunding: New but may be worth investigating.
- **Form multiple banking relationships:** I recommend establishing relations with two financial institutions: one should be a credit union and the other should be either a commercial bank or a savings bank.
- **Lending authority:** A local branch manager has less "approval authority" than does a senior manager at the main office.

- **SBA:** The SBA does not give loans; it guarantees a portion of loans. Also, it does not offer grants to start or grow a business. For more information, visit: www.sba.gov.
- **Due diligence:** The financial institution that you go to is required to perform what is called "due diligence" and will require much documentation from you.
- **Be prepared:** It is important that you prepare for your meeting with the loan officer.
  - Bring your Business Plan and Net-Worth Statement.
  - Be prepared to discuss your credit rating, background and social media postings.
  - Dress Professionally.
  - Be prepared to present a 20 minute presentation of your Business Plan.
  - Thank the loan officer for his time and consideration.
  - A "thank you" note or e-mail is always a good idea.
- **Security:** Lenders will want some type of security for their loan. Possibilities include guarantor, endorser, co-signer, demand note, or mortgage.
- **Receivables financing:** Funds are often needed to fund operations while waiting for your clients to pay.

CHAPTER 6

BUSINESS PLAN

INTRODUCTION

You will understand what a Business Plan includes, how you can build one, and how important it will be to the success of your engineering practice. Discover how your business plan defines your company and its goals, gives you a game plan, and helps you make a go/no go/or wait decision. Find out why updating your business plan both monthly and annually allows you to adjust expectations and/or modify actions to compensate. Observe that your business plan will be the "key tool" in seeking financing and a line of credit, both for start-up and for cash flow needs.

WHAT IS A BUSINESS PLAN?

A "Business Plan" is a formal statement of business goals, reasons they are attainable, and plans for reaching them. It should also contain background information about the organization attempting to reach those goals.

WHY IS A BUSINESS PLAN IMPORTANT?

- It is the "structure" of your practice which precisely defines your company and identifies your goals.
- It will organize your thoughts and will give you a "game plan."

- It will help you make a "go/no go/or wait" decision.
- It will help you in forecasting funding needs and securing "financing."

WRITE A BUSINESS PLAN

This is one of the most important and frequently overlooked steps in the process of starting your own engineering practice. *When I started my practice, I didn't even know what a business plan was. That is probably why I "stumbled" along by the seat of my pants and succeeded by luck and perseverance.*

It is critically important that you have patience and take the time to properly research and thoughtfully write a clear and concise document. It will be necessary to update this business "guide" annually.

During your first year in business, you will find it helpful to compare projections to actual cash flows on a monthly basis, to adjust expectations accordingly, or to modify actions to compensate.

WHAT DOES A BUSINESS PLAN INCLUDE?

- Cover Page
- Table of Contents
- Executive Summary
- Business Description
- My Qualifications (or "The Management Team")
- Ownership Agreement (if not a Sole Proprietor)
- Operations Plan
- Legal, Records/Taxes & Insurance
- Marketing Plans
- Cash Outflow Projections
- Cash Inflow Projections
- Cash Flow Analysis
- Balance Sheets
- Financing Overview & Funding Needs
- Supplemental Materials

OUTLINE SUMMARY

- **Be careful and realistic:** A "Business Plan" is one of the most important and frequently overlooked steps in the process of starting an engineering practice.
- **Make it your game plan:** Your Business Plan defines your company and its goals, helps you (and your spouse/partners) make a "go/no go/or wait" decision, and aids in forecasting financial needs and securing financing.
- **Update:** Look at your Business Plan both monthly and annually. Compare projections to actual and adjust expectations accordingly or modify actions to compensate.
- **Company name:** Carefully choose your company name considering both how it identifies you and the future salability of your practice.
- **Key tool:** Your Business Plan will be critically important in seeking financial support, financing, and a line of credit for start-up and cash flow needs.

PART 3

IMPLEMENTATION

CHAPTER 7

CASTING OFF (STARTING YOUR OWN PRACTICE)

INTRODUCTION

You will discover why checklists are your friends: they remind you of all the necessary steps and help you avoid the missed steps. The "Pre-Start-Up Checklist" will reveal the many necessary and time consuming details that can be done in advance. The "Final Start-Up Checklist" will uncover which things should only be done after you have Cast Off.

GENERAL

"Casting Off" is like starting on a long sea journey; make sure that your vessel is sea worthy. Be prepared for the voyage by reviewing the "Pre-Start-Up Checklist" and the "Final Start-Up Checklist."

I assure you that if you have followed my advice, this will be the trip of your life and lead to an exceptional engineering practice, career satisfaction, and financial security. *Bon voyage* and calm seas.

OUTLINE SUMMARY

- **"PRE-START-UP CHECKLIST"**
  - *Verify "Ownership Agreement"
  - *Verify "Cash Availability"
  - *Apply for a "Loan" and/or a "Line of Credit" (if needed)
  - Prepare a realistic "Time Table" for all activities
  - Update insurance quotes
  - Prepare your "Company Brochure"
  - Prepare your "Business Announcement"
  - Select your "Letterhead & Business Cards"
  - Plan your "Website"
  - Update your "Client List"
  - Make a list of "Library Materials" needed (catalogs, etc.)
  - Prepare your "Standard Documents"
  - Prepare your "Operations Manual"
  - Prepare "CAD & BIM Files" and standards
  - Finalize "Rental Space" selection

- **"FINAL START-UP CHECKLIST".**
  (Do only after the first three items in the "PRE-START-UP CHECKLIST" are completed and approved.)
  - "Rent/Lease" office space
  - Deal with "legal matters"
  - Obtain "General Insurance"
  - Obtain "Professional Liability Insurance"
  - Meet with your "CPA"
  - Order check writing system or accounting software
  - Complete Office "Fit-Out"
  - Purchase & Installation of Office "Phone System"
  - Buy/Rent "Office Furniture"
  - Buy/Rent "Office Equipment"
  - Buy computer software (including CAD & BIM)
  - Buy "Office Supplies"

- Print "Company Brochure"
- Print "Business Announcement"
- Print your "Letterhead & Business Cards"
- Hire a website designer
- "Move in & Setup"

- **BE PREPARED**
  - "By failing to prepare, you are preparing to fail."

PART 4

MANAGING YOUR FIRM

CHAPTER 8

LEGAL, ACCOUNTING & INSURANCE

INTRODUCTION

Legal

You will learn that the "business entity" that you choose for your engineering practice will have major legal and tax implications. Discover which government agency will help you to choose the best structure for your situation. Find out the special liability issues that PEs have of which some attorneys are unaware. Detect concerns that are often overlooked when choosing a business structure. Uncover how to save money using common sense.

Accounting

You will realize that all accountants are not CPAs and why your firm should hire a CPA. Discern the "tax" implications of your "business structure" and which method of accounting, "cash" or "accrual," usually works better for a small professional practice. Learn how careful "time tracking" will benefit your practice by uncovering inaccuracies and inefficiencies. Understand when to progress from "manual" time tracking to "off-the-shelf" solutions and then to "customized" accounting software.

Insurance

Hear about what type of "general" insurance all consulting engineering firms need and which may be optional. Understand why you need "professional liability/errors and omissions" insurance, what is meant by a "claims-made" basis, and when you need "retroactive" coverage. Find out which employee benefit you can't overlook and the percent the employee might be expected to contribute.

PROFESSIONAL LIABILITY INSURERS

ACE USA Insurance
601 So. Figueroa St., 15th Fl.
Los Angeles, CA 90017
(213) 833-3100

AEIC
2056 Westings Ave., Suite 20
Naperville, IL 60563
(800) 437-2342

Argo Pro Insurance
2000 S. Batavia St.
Geneva, IL 60134
(800) 447-4626

Arrowhead Design Insurance
99 Pacific St., Suite 155F
Monterey, CA 93940
(831) 333-9840

Aspen Insurance
590 Madison Ave., 7th Fl.
New York, NY 10022
(646) 502-1012

AXIS Insurance
300 Connell Dr., Suite 8000
Berkeley Heights, NJ 07922
(908) 673-2963

Beazley Insurance
141 Tremont St., Suite 1200
Boston, MA 02111
(617) 239-2600

Catlin Design Professional
1331 N. California Blvd.
Walnut Creek, CA 94596
(925) 927-2239

CNA/ Schinnerer & Co.
2 Wisconsin Circle
Chevy Chase, MD 20815
(301) 961-9800

Euclid Managers Insurance
234 Spring Lake Dr.
Itasca, IL 60143
(630) 238-1900

Hanover Insurance Group
333 W. Pierce Rd., Suite 300
Itasca, IL 60143
(630) 760-3045

HCC Specialty Insurance
2300 Clayton Rd., Suite 1100
Concord CA 94520
(925) 685-1600

James River Insurance
PO Box 27648
Richmond, VA 23261
(804) 289-2700

Lexington Insurance Co.
100 Summer St.
Boston, MA 02110
(617) 330-8564

Liberty Int. Underwriters
LIU USA, 18th Fl.
55 Water St.
New York, NY 10041
(212) 208-4100

Markel Insurance
10 Parkway North
Deerfield, IL 60015
(847) 572-6000

Navigators Management Co.
6011 University Blvd., St. 280
Ellicott City, MD 21043
(443) 364-5938

RLI Insurance Co.
150 Monument Rd., St. 605
Bala Cynwyd, PA 19004
(610) 664-8700

Travelers Insurance
111 Schilling Rd.
Hunt Valley, MD 21031
(443) 353-2263

XL Insurance
30 Ragsdale Dr., Suite 201
Monterey CA 93940
(800) 227-8533

FIGURE 8-1

From the start of my practice, I insured with the CNA/Schinnerer Professional Liability Insurance Program, and I stayed with them for my tail coverage when I sold my practice. They have been partnered with both NSPE and AIA for over 40 years. Go to www.shinnerer.com, www.nspe.org, and www.aia.org to check them out.

I stayed with CNA/Schinnerer primarily for their "pre-claims assistance" and their "guidelines." In my over 35 years of practice, I had a few situations that I thought might result in a lawsuit. I contacted the insurer (through my insurance broker), and they connected me with a great local lawyer who specialized in construction claims. I met with the lawyer two or three times and, fortunately, the suits never materialized. There was no charge to me, and my insurance premiums did not increase.

Four times a year, CNA/Schinnerer produces "guidelines," a four page document with short articles on improving your practice. I ended up with a thick binder of these and almost always found something, in every issue, to share with my staff.

OUTLINE SUMMARY

- **Business structure:** The "business structure" that you choose for your engineering practice will have major legal and tax implications.
- **SBA:** The Small Business Administration's website (www.sba.gov) has helpful articles describing the major types of business entities.
- **Legal:** Seek a recommendation to a small law firm and ask for an initial meeting; don't be afraid to enquire about the cost.
- **Personal liability:** Make sure that your attorney is aware that each PE who stamps/seals documents cannot avoid personal liability for his errors and omissions by having a corporate business structure.
- **Considerations:** Don't forget to consider your CPA's input, same or similar company names in use, availability of web address/URL, or "fictitious name" filing.
- **Evaluate legal needs:** Don't hesitate to consult your attorney for important legal matters, but avoid costly consultations on trivial things.
- **CPA:** Look for a small CPA firm (not just an accountant) that regularly handles "start-up work" for small businesses.
- **Accounting method:** The "cash" method of accounting, rather than the "accrual" method, usually works better regarding taxes for a small professional practice.
- **Track time:** An important aspect of your business is carefully keeping track of time; it is the only thing that you have to sell. Tracking time will help you to make more accurate budgets, discover inefficiencies, identify difficult clients, and uncover "scope creep."
- **Simple time tracking:** Transfer time card hours to project summary sheets and summarize monthly. Manual tracking of time will be easier and more accurate using computer spread sheets.

- **Off-the-shelf software:** As your firm expands, consider getting "off-the-shelf" accounting software; QuickBooks and Sage 50 are popular choices.
- **Customizable software:** If you have special billing needs or when you grow larger, it may be worth investigating "customizable" accounting software. The Accounts Receivable Aging report and Budgeted vs. Actual hours are most useful.
- **Insurance:** You need to be the "risk manager" for your company when purchasing insurance.
- **Insurance needs:** General insurance needs for each professional practice can be different but usually include comprehensive, auto, and fire and theft.
- **Errors & omissions:** Engineers always try to do their best work, but mistakes sometimes occur, resulting in lawsuits. Professional liability insurance (PLI), more commonly known as errors and omissions (E&O), is a form of liability insurance that helps protect professionals.
- **Claims made:** PLI policies are generally setup on a "claims-made" basis, meaning that the policy only covers claims made during the policy period. If you change insurance companies, make sure that you are covered retroactively. I encourage you to consider both coverage and services provided, in addition to price.
- **Group health insurance:** Health insurance is the one area that you can't initially overlook. For a "start-up" company, consider having employees contribute 30% for themselves and 40% for their families.

CHAPTER 9

MARKETING MATERIALS

INTRODUCTION

Find out how to prepare your business announcement, letterhead, business cards, and brochure distinct from those of your competitors. Learn how to make your mailing list one of your most important promotional tools and why to include more than just existing and potential clients. Discover the lists and forms that your new practice will need, and discern a simple way to avoid construction headaches. Most importantly, understand the secret to gaining a reputation as an "expert" in your field.

GENERAL

I think that the two chapters on MARKETING are the <u>most</u> important chapters in this book. Finding and keeping clients is the key to success. Most people, especially engineers, are "shy." Get over it!

While attending college, I worked two summers selling encyclopedias door-to-door. In the process, I gained confidence.

I am not suggesting a part-time job, but you do need to get over your timidity. Your activities described under "Get Know & Gain Credentials" in Chapter 2 should have provided many opportunities to meet new people and occasions for public speaking. If you followed my advice, you are ready for marketing.

The following are some documents to prepare once you have made the decision to open your own practice. But don't actually get them printed until you are ready to "Cast Off" (See Chapter 7).

Your Brochure

Your brochure may evolve over time. But, most importantly, make it something of which you are "proud."

My "fourth" was a "mini-brochure": a single 8 ½ x 14 parchment sheet folded in thirds. It was only used as a handout at events, like speaking engagements with numerous attendees, when it was too expensive to provide my normal brochure. See Figure **9-1**.

BROCHURE (1 OF 2)

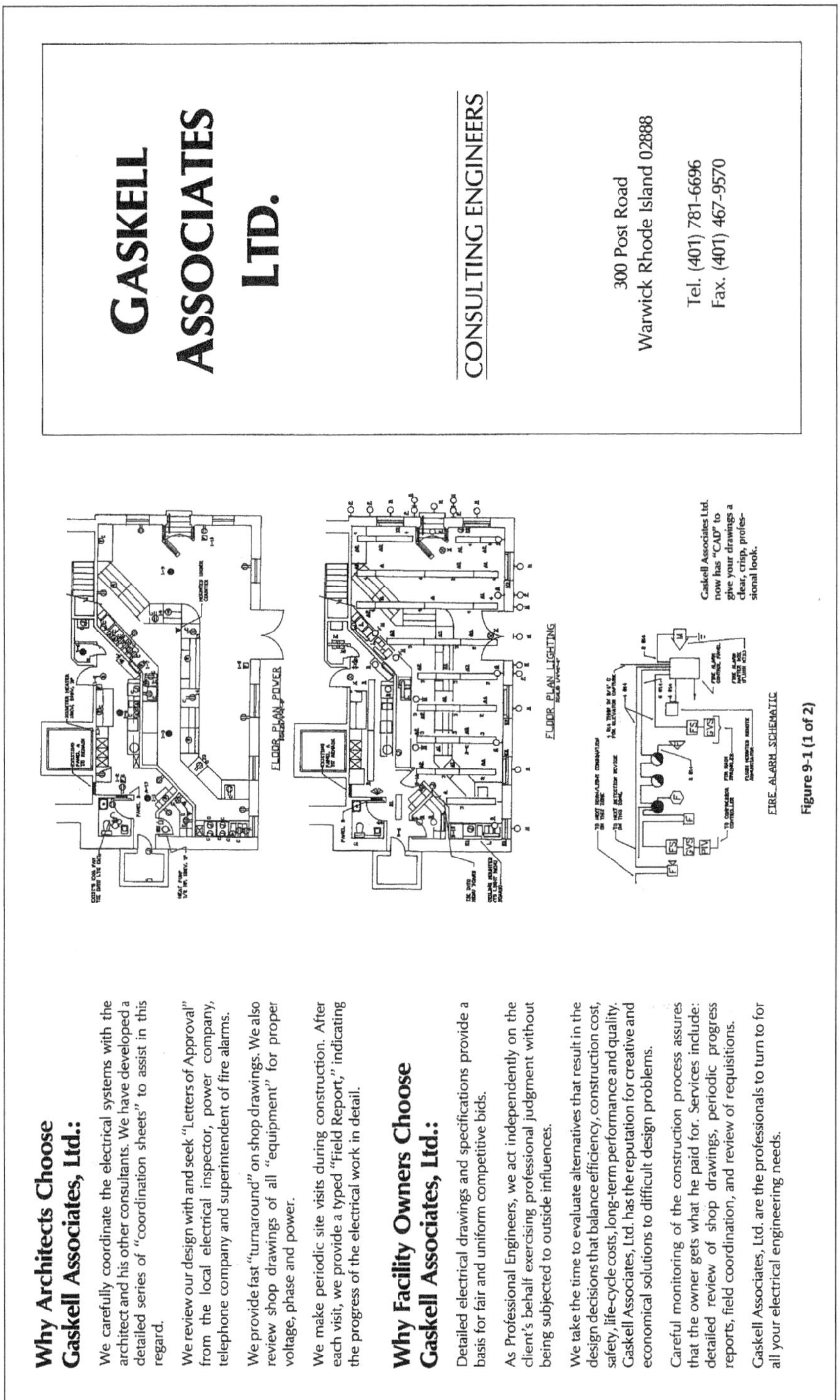

Why Architects Choose Gaskell Associates, Ltd.:

We carefully coordinate the electrical systems with the architect and his other consultants. We have developed a detailed series of "coordination sheets" to assist in this regard.

We review our design with and seek "Letters of Approval" from the local electrical inspector, power company, telephone company and superintendent of fire alarms.

We provide fast "turnaround" on shop drawings. We also review shop drawings of all "equipment" for proper voltage, phase and power.

We make periodic site visits during construction. After each visit, we provide a typed "Field Report," indicating the progress of the electrical work in detail.

Why Facility Owners Choose Gaskell Associates, Ltd.:

Detailed electrical drawings and specifications provide a basis for fair and uniform competitive bids.

As Professional Engineers, we act independently on the client's behalf exercising professional judgment without being subjected to outside influences.

We take the time to evaluate alternatives that result in the design decisions that balance efficiency, construction cost, safety, life-cycle costs, long-term performance and quality. Gaskell Associates, Ltd. has the reputation for creative and economical solutions to difficult design problems.

Careful monitoring of the construction process assures that the owner gets what he paid for. Services include: detailed review of shop drawings, periodic progress reports, field coordination, and review of requisitions.

Gaskell Associates, Ltd. are the professionals to turn to for all your electrical engineering needs.

FLOOR PLAN POWER

FLOOR PLAN LIGHTING

FIRE ALARM SCHEMATIC

Gaskell Associates Ltd. now has "CAD" to give your drawings a clear, crisp, professional look.

Figure 9-1 (1 of 2)

GASKELL ASSOCIATES LTD.

CONSULTING ENGINEERS

300 Post Road
Warwick Rhode Island 02888

Tel. (401) 781-6696
Fax. (401) 467-9570

FIGURE 9-1

BROCHURE (2 OF 2)

Gaskell Associates, Ltd.

Gaskell Associates, Ltd. is a consulting engineering firm primarily involved with the design of *electrical systems for buildings.*

Our specialties include:

- Power Distribution Systems
- Emergency Power Systems
- High Voltage Power Distribution Systems
- Lighting Design (Interior, Exterior, Area and Roadway)
- Fire Alarm Systems
- Burglar Alarm Systems
- Communications Systems
- MATV and CCTV Systems
- Computer Power and Network Systems
- UPS and Power Conditioning Systems
- Feasibility Studies

In addition, we have developed a team of consultants to allow us to offer a full range of incidental services including: HVAC, plumbing, architectural, structural, and civil engineering.

Life-cycle-costing is a major element of our design effort and energy conservation is always considered. This results in a facility that is economical, efficient, functional, and easily maintained, as well as one that is aesthetically pleasing.

Gaskell Associates, Ltd. endeavors, at all times, to maintain its reputation for accuracy, attention to detail, and punctuality. We have executed engineering projects for almost every type of facility, including numerous renovations.

This engineering practice was established in 1971 by John D. Gaskell, P.E., electrical engineer. We now employ a staff of eleven.

John D. Gaskell, P.E.

Gaskell Associates, Ltd. is headed by John D. Gaskell, P.E. Jack is a graduate of Wentworth Institute, with an Associates degree in Electrical Engineering Technology, and the University of Rhode Island, with a Bachelor of Science degree in Electrical Engineering. He is a Registered Professional Engineer in Rhode Island, Massachusetts, Connecticut, New Hampshire, New York, New Jersey, Pennsylvania, and Florida.

Jack is a member of the following professional organizations: Rhode Island Society of Professional Engineers, Providence Engineering Society, International Association of Electrical Inspectors, Electrical League of Rhode Island, Illuminating Engineering Society, and the Institute of Electrical and Electronic Engineers.

He is a past president of the Rhdoe Island Society of Professional Engineers; past president of the Rhode Island Chapter of the Illuminating Engineering Society, past director of the Electrical League of Rhode Island; and past director of the Rhode Island Chapter of the International Association of Electrical Inspectors.

Jack writes the electrical specifications for our projects and provides executive project management. He has written several articles for national technical publications. His design of a one megawatt Uninterruptible Power System for Fleet National Bank's Operations Center won "first place" in a national design award competition.

Jack was recently honored by being selected "Engineer of the Year" by the Rhode Island Society of Professional Engineers.

Recent Clients

a i designs, inc., Pawtucket, Rhode Island
Gordon R. Archibald, Inc., Pawtucket, Rhode Island
Arris Design, Inc., Providence, Rhode Island
James Barnes, AIA, Providence, Rhode Island
Benefit Street Design, Inc., Providence, Rhode Island
Brown University, Providence, Rhode Island
Citizens Bank, Properties Dept., Providence, Rhode Island
The Coken Company, Providence, Rhode Island
Corporate Concepts, Warwick,, Rhode Island
Cranston Housing Authority, Cranston, Rhode Island
Daughn/Salisbury, Inc., Providence, Rhode Island
Diocese of Providence, Providence, Rhode Island
Di Leonardo International, Inc., Warwick, Rhode Island
East Providence Housing Authority, East Providence, Rhode Island
Ekman, Arp and Snider, Warwick, Rhode Island
Charles B. Fink, A.I.A., Providence, Rhode Island
Fleet National Bank, Properties Dept., Providence, Rhode Island
Garofalo Associates, Inc., Warwick, Rhode Island
Gates, Leighton & Associates, Inc., East Providence, Rhode Island
Gilbert & Maloney, Providence, Rhode Island
Irving B. Haynes & Associates, Providence, Rhode Island
The Hillier Group, Princeton, New Jersey
Kent Cruise & Partners, Providence, Rhode Island
Lamborghini and Feibelman, Providence, Rhode Island
Lincoln Housing Authority, Lincoln, Rhode Island
Long, Staats and Associates, Newport, Rhode Island
Americo Mallozzi, A.I.A., Providence, Rhode Island
George Morin Architects, Inc., Woonsocket, Rhode Island
Kevin S. Munroe - Architect, Wakefield, Rhode Island
Naval Underwater Systems Center, Newport, Rhode Island
O'Hearne Associates, Woonsocket, Rhode Island
Lee Pare & Associates, Inc., Providence, Rhode Island
City Of Pawtucket School Department, Pawtucket, Rhode Island
PPG Industries, Pittsburgh, Pennsylvania
Lombard J. Pozzi, A.I.A., Warren, Rhode Island
The Robinson Green Beretta Corp., (RGB), Providence, Rhode Island
Robinson, Myrick & Associates, Inc., Smithfield, Rhode Island
Roger Williams General Hospital, Providence, Rhode Island
The Ritchie Organization (TRO), Newton, Massachusetts
Rhode Island Department of Environmental Management
Rhode Island Department of Transportation
Rhode Island Division of Water Resources
Norton E. Salk, Architect, Cranston, Rhode Island
Raymond W. Schwab Associates, Inc., Peacedale, Rhode Island
St. Joseph Hospital, Fatima Unit, N. Providence, Rhode Island
St. Joseph Hospital, Providence Unit, Providence, Rhode Island
United States Navy, Newport, Rhode Island
Technical Materials, Inc., Lincoln, Rhode Island
Warren Housing Authority, Warren, Rhode Island
Warwick Housing Authority, Warwick, Rhode Island
City of Woonsocket, Planning Department, Woonsocket, RI
Womens Development Center, Providence, Rhode Island
Veteran's Medical Center, Providence, Rhode Island

Figure 9-1 (2 of 2)

FIGURE 9-1

Reprints of Articles

During my second or third year in business, power companies started to reduce their voltage to conserve energy. Clients became concerned that this might have a deleterious effect on their motors, lights, and electrical systems. I did some reading of textbooks and handbooks, made some calculations to quantify the results, and included this information in an update newsletter. It got rave reviews, so I expanded this information into a magazine article that actually got published. I ordered 1,000 reprints from the magazine and sent some of them to my mailing list. I used the extras as filler to my brochure. Overnight, I was thought to be an "expert."

The following is an example of a reprint of one of my articles:

- Article about a technical issue, "Electrical Design with EMF in Mind," written by me and published in EC&M. It was 11 x 17 folded sheet. See Figure **9-2.**

EC&M ARTICLE (1 OF 3)

EC&M

ELECTRICAL CONSTRUCTION AND MAINTENANCE

Electrical Design With EMF In Mind

What are electromagnetic fields and what efforts should be made to minimize them in designing electrical systems today?

ELECTRICAL ENGINEERING BY
GASKELL ASSOCIATES, LTD.
CONSULTING ENGINEERS

Figure 9-2 (1 of 3)

FIGURE 9-2

EC&M ARTICLE (2 OF 3)

WHAT'S THE STORY

Electrical Design With EMF In Mind

What are electromagnetic fields and what efforts should be made to minimize them in designing electrical systems today?

John D. Gaskell, P.E.
John D. Gaskell, P.E. is President of Gaskell Associates, Ltd., a consulting engineering firm in Warwick, R.I., and a member of the National Electromagnetic Field Testing Association.

While it may be years before a clear consensus is established on the health risks of electromagnetic fields (EMF) exposure, the topic remains highly controversial. In the meantime, electrical systems can be designed and installed to minimize the extent of the magnetic field component of EMF, often at little or no added cost. Before discussing the steps that can be taken during the design stages of electrical distribution systems, let's find out what EMF is.

What is EMF?

EMF at 60Hz is really made up of two separate entities: electric fields and magnetic fields. An *electric* field, which exists when voltage is present and which is easily blocked by metal, can cause currents to flow on the *surface* of the human body. Electric fields are *not* generally considered to be a biological hazard.

A *magnetic* field, which exists when current flows and which is *not* appreciably blocked by common materials, can cause current to flow *through* the human body. Magnetic fields at 60 Hz *are* considered a possible biological hazard.

A magnetic field decreases as the distance from the source increases. However, the configuration of the source actually determines how quickly a field diminishes. Multiple conductors with current flowing in opposite directions or 3-phase circuits have magnetic fields that are inversely proportional to the distance *squared*. Appliances and transformers are point sources and their fields drops off inversely proportional to the distance *cubed*.

Exposure factors

Without consensus in the scientific community that a hazard actually exists, economics play a big factor in setting limits. We don't know for sure whether long-term, low-level exposure is worse than short-term, high-level exposure.

Some say that switched fields are more dangerous than steady-state fields. Individual characteristics of the person being exposed may need to be considered; age, health, and even fertility and pregnancy may be factors. Obviously, more research is needed. But in the meantime, it may be practical to make some educated guesses, set some *interim* exposure limits, and consider some changes in design.

Practical design suggestions

Services. For an overhead MV service lateral, consider selecting spacer cable in lieu of cross-arm construction. Utilities are becoming more sensitive to the magnetic field issue and may welcome your suggestions. Another choice for a service lateral, of course, is an underground service.

Preferably, pass a service lateral under a storage room rather than an office that is occupied for long time periods. If this cannot be done, then the service lateral should be enclosed in a metallic raceway. The magnetic field will induce a counter EMF in the metallic raceway, which will help reduce the field.

Locate pad-mounted transformers at least 20 ft from buildings. Another consideration is to encircle a pad-mounted transformer with a fence, located at a 4- to 6-ft distance from the edge of the pad.

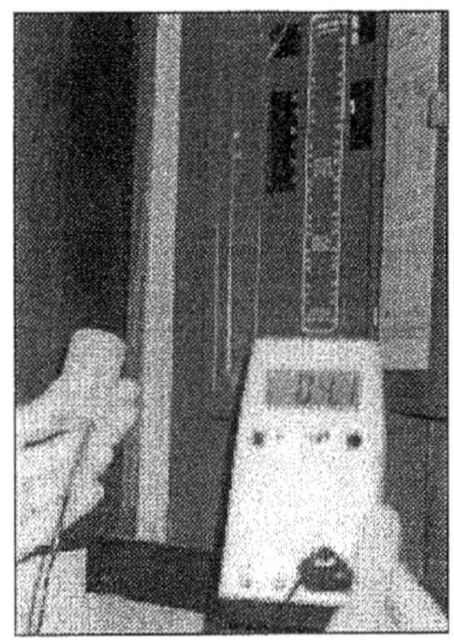

A magnetic field meter assembly consists of a digital multimeter, modified for additional shielding, and a magnetic sensor.

Switchboards and panels. Locate distribution equipment on exterior walls or walls adjacent to storage areas or corridors. Where possible, locate free-standing switchboards to allow 3 ft of working space on all sides. Use walls abutting occupied spaces for telephone or fire alarm equipment. Avoid locating an electrical room directly below or above an occupied space.

Indoor transformers. Although a typical dry-type, step-down power transformer creates a large magnetic field, this field's strength dissipates very quickly. However, it would still be good practice to locate such a unit in an electrical room remote from occupied spaces.

Bus duct. For vertical distribution of power in a tall building, follow a procedure similar to an indoor transformer siting. Where possible, run a vertical bus duct on a wall common to the elevator shaft, janitor closet, or corridor. In a factory, a bus duct run should be located away from operator positions.

Underfloor ducts. Individual conductors are usually installed in an underfloor duct system, and it is possible for the phase and

EC&M February 1995

Figure 9-2 (2 of 3)

FIGURE 9-2

EC&M ARTICLE (3 OF 3)

WHAT'S THE STORY

Electrical Design With EMF In Mind

neutral conductors of a 2-wire branch circuit to be from 6 to 23 in. apart (depending on the cross section of the cell). This configuration would create significant magnetic fields. Recommended practice would be to twist the individual conductors of a branch circuit together in pairs. For a large new construction project, twisted cable assemblies can be purchased from specialty cable companies with only a slight increase in cost. At an existing installation, all the wires in the duct could be tie-wrapped at each outlet in the duct system.

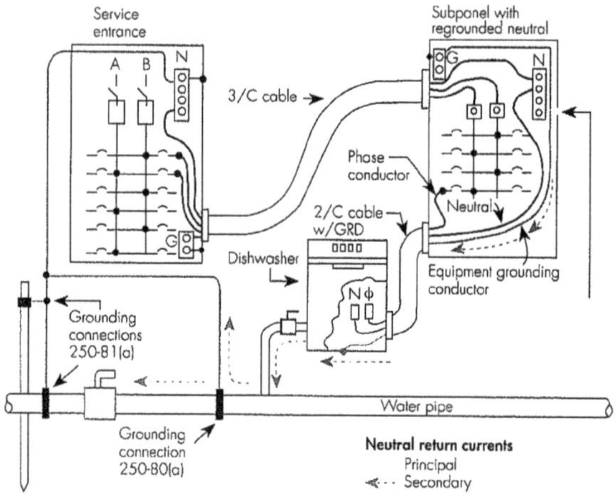

An example of a connection that allows a water pipe to function as a parallel neutral. The equipment grounding conductor run to the neutral bus of the subpanel is a violation of the NEC. This type of wiring error is widespread.

What to watch out for

Wiring errors. When conductors are installed in such a way that circuit conductors are not in the same cable or raceway, substantial magnetic fields are created by the separated supply and return currents. The most common error is violating Sec. 250-23(a) by making a connection between a neutral and an equipment grounding conductor on the load side of the service disconnect; this usually happens at sub-panels. In this instance, neutral currents will flow on *both* the neutral and the equipment grounding conductor.

Other errors include incorrect wiring of 3-way switching circuits and the connection of neutrals from two different branch circuits at some point other than panel neutral busses, such as at junction boxes, switches, receptacles, etc.

Water pipe problems. Three specific areas can be addressed here.

• Neutral grounds to a water pipe on the load side of the service can cause problems similar to the case mentioned above, where Sec. 250-23(a) is violated by making a connection between a neutral and an equipment grounding conductor, on the load side of the service disconnect. If any electrical equipment, such as a hot water heater, is also connected to a metal water pipe, the water pipe will probably become a parallel path for current that should be flowing over the neutral.

• Physical damage or corrosion can cause the neutral conductor (often uninsulated) on an overhead service drop to open up, or sever. With this condition and with water supply laterals connected to conductive water mains, the water system can function as a parallel neutral. When this happens, unbalanced neutral currents seeking to return to the utility source will pass out of the building over the water lateral. They will return to the source after passing through the water main, a neighboring water lateral, and then through the neighboring service disconnect and out over the neighboring neutral.

• Currents can originate outside a building from the power company distribution system (typically, a looped primary circuit) or from neighboring buildings via the water service. An insulating coupling, or fitting, can be installed in the water service. However, it must be located *outside* of the premises, at least 10 ft from the building wall, per NEC Sec. 250-81(a).

Note that installing an insulating fitting in a water line to prevent the entry of currents has the disadvantage of decreasing ground conductivity, which under certain fault conditions may cause excess voltage to appear on plumbing fixtures and appliance enclosures. Thus, an insulating fitting can cause both a shock hazard and damage to an appliance that would not have occurred if there had been a connection to the water system.

However, a device called an automatic ground connector, which reconnects the ground in the event of a severe fault, is available. This device is similar in principle to a conventional surge arrester, but operates on a different voltage range. Because this is a new device that has not yet been tested by a recognized testing authority, approval from the local authority having jurisdiction should be obtained.

When an insulating fitting is installed, a ground wire should be clamped beyond the fitting and extended back into the building. This allows the option of installing an automatic ground connector or reestablishing the ground in the future.

The importance of testing

Conducting a magnetic field survey upon completion of an installation in new construction is recommended. Such a survey can be done also at an existing location to detect long standing wiring errors or other abnormal conditions.

Two different types of meters are used. A single coil, or single axis, meter (about $235) determines the direction of the field and is useful for finding the field source. This type of meter can provide an approximate rms reading, but a calculation is required.

A 3-coil or 3-axis meter (about $1400) gives a true rms reading of fields from all directions and is not dependent on the orientation of the meter. Capable of making readings every 4 sec over a 24-hr period, this type of meter can be left at a site, or a person can carry it along a predetermined path. Accumulated data can be downloaded into a computer. •

Figure 9-2 (3 of 3)

FIGURE 9-2

Award Announcements

Anytime someone in your firm receives a noteworthy award, send out an announcement.

The following is an examples:

- Announcement of award, "I.E.S. Lighting Design Award for the Hall Library Lighting Design," by the New England Section of IES. See Figure **9-3.**

AWARD ANNOUNCEMENT

I.E.S. AWARD PRESENTED FOR HALL LIBRARY LIGHTING DESIGN

The restoration of the Hall Library in Cranston, Rhode Island called for a complete lighting upgrade to assure footcandle levels consistent with modern standards. In keeping with the historic nature of the building, existing fixtures were retrofitted with high lumen H.I.D. sources and supplemented with additional luminairs to produce award winning results.

Architect	The Robinson Green Beretta Corp.
Architectural Project Manager	Frederik R. Love, A.I.A.
Electrical Project Manager	Jefferey L. Clymer

Jack Gaskell, President of Gaskell Associates, Ltd. with Jeff Clymer, Project Manager at IESNA New England Section Awards Dinner.

Chain-hung globe-type fixtures were retrofitted to metal halide and supplemented with new recessed metal halide down lights.

False skylite was back lighted by indirect metal halide floods. Half inch parabolic cubed louvers control glare from fluorescent fixtures at main desk.

Fluorescent fixtures provide excellent vertical footcandle levels on book stacks. Up-light from globes highlight the hand carved wood moldings.

GASKELL ASSOCIATES, LTD. *CONSULTING ENGINEERS*

300 Post Road, Warwick, RI 02888 (401) 781-6696

Figure 9-3 (1 of 1)

FIGURE 9-3

Announcements of New Services or New Specialties

Whenever you start a new service or new specialty, send out an announcement.

The following is an example:

- Announcement of new service, "Power Quality Monitoring." See Figure **9-4.**

Send these to your mailing list and have extra copies made to be used as brochure filler. Also, place a (free) announcement along with a picture in the local Sunday newspaper. Also, send it to related organizations and suggest that they make mention of it in their newsletters.

ANNOUNCEMENT OF NEW SERVICE (1 OF 2)

Power Quality Monitoring and Certification Service is a subsidiary of Gaskell Associates, Ltd. Electrical Consulting Engineers. Established in 1971, they are the largest electrical engineering firm in Rhode Island.

Gaskell Associates, Ltd. is headed by John D. Gaskell, P.E. Jack is a graduate of the University of Rhode Island, with a Bachelor of Science degree in Electrical Engineering. He is a Registered Professional Engineer in Rhode Island, Massachusetts, Connecticut, New York, New Jersey, and New Hampshire.

Jack is a member of the following professional organizations: Rhode Island Society of Professional Engineers, Providence Engineering Society, Electrical League of Rhode Island, Illuminating Engineering Society, National Electromagnetic Field Testing Association, National Forensic Center, and the National Academy of Forensic Engineers.

He is past President of the Rhode Island Society of Professional Engineers and a past President of the Rhode Island Chapter of the Illuminating Engineering Society. Jack is currently a Director of the Electrical League of Rhode Island and is a member of both the Engineering Advisory Board and the Corporation of Roger Williams University. He is also a member of the Rhode Island Building Codes Standards Committee.

Jack has written numerous articles for national technical publications and has been a guest speaker at the National Conference on Harmonics and Power Quality. He is a nationally recognized expert on magnetic fields. Jack was recently honored by being selected "Engineer of the Year" by the Rhode Island Society of Professional Engineers and is recipient of the Providence Engineering Society's "Freeman Award" for advancement in the art and practice of engineering.

Call Today

Our service is intended to save money by extending equipment life and improving the reliability of the equipment and facility that your business depends upon.

The fee for our services depends upon the sophistication of the equipment we provide and the complexity and frequency of reports that you require.

We will be happy to customize a no obligation quotation for our services to suit your facility, match your needs, and meet your budget.

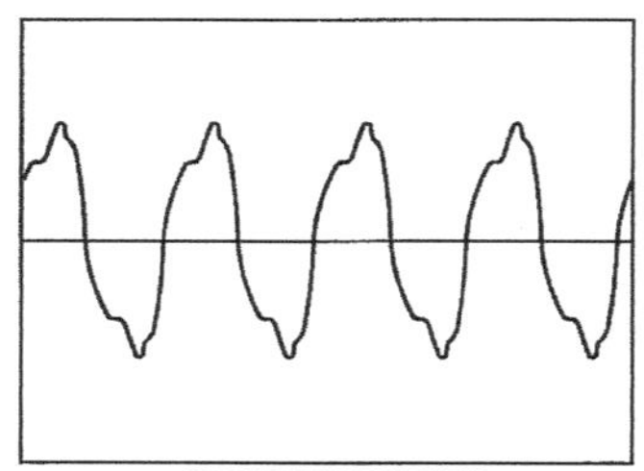

Figure 9-4 (1 of 2)

Power Quality Monitoring

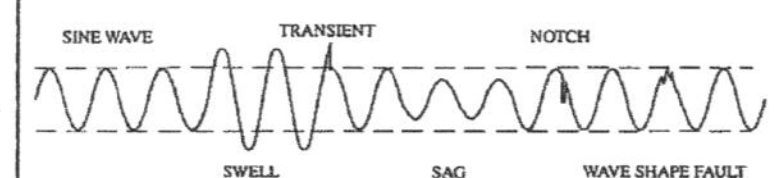

POWER QUALITY MONITORING

& CERTIFICATION SERVICES

a division of

GASKELL ASSOCIATES, LTD.

Consulting Electrical Engineers

300 POST ROAD

WARWICK RHODE ISLAND 02888

Tel. (401) 781-6696

Fax (401) 467-9570

Toll Free (888) 781-6696

FIGURE 9-4

ANNOUNCEMENT OF NEW SERVICE (2 OF 2)

Summary:

For a small monthly fee, **"Power Quality Monitoring & Certification Services"** will provide equipment to continuously monitor power quality at your facility and provide periodic reports summarizing and analyzing data recorded. Detecting and correcting power quality problems early may avoid damage to equipment or loss of critical data. Proof of acceptable power quality, (certified by a **Professional Engineer**) may prevent "voiding" of your equipment warranties/guarantees. Proof that power anomalies originated on the power company system may qualify you for reimbursement if damage occurs.

Our Service Offers Two Unique Advantages:

1. There is no need for you to purchase equipment, pay for installation and wiring, or hire a specialist to interpret the recordings.

2. Our reports are "certified" by a licensed, experienced, independent **Professional Engineer***.

*Note: John D. Gaskell, P.E. is a Registered Professional Engineer in RI, MA, CT, NH, NY, & NJ

How This Service Works:

Basically, our power quality monitoring service works as follows:

We install one or more sophisticated power quality monitors at your facility. There is no cost to you for equipment, installation* and associated wiring and maintenance. *(*Note: Beyond 100 miles from R.I., there is a small one-time installation charge).*

Monitoring capabilities depend upon the "LEVEL" of the equipment installed. Level I equipment will continuously monitor and periodically record current and voltage on all phases and neutral, as well as voltage between neutral and ground. In addition, it will detect and record the following power anomalies:

•Transients	•Overvoltages
•Sags	•Undervoltages
•Swells	•Interruptions

We provide a telephone modem and connect to a telephone line which you extend to the vicinity of each meter and assign for our use. We periodically call your facility and download the data recorded.

We prepare reports (usually monthly) which summarize and analyze the data accumulated.

Why Your Facility Should Be Monitored.

Facilities now have computer based loads that are more sensitive to the imperfections that have always existed in the power supplied to them. Power aberrations can cause reduced equipment/component life, software damage, memory loss, equipment failures and power outages. With power company deregulation and assosciated cost cutting, many business owners are concerned that they will be unaware of a decline in power quality, until damage has already occurred.

Our reports, certified by a **Professional Engineer**, usually provide proof that your power quality for the period was within industry accepted tolerances. This proof may prevent your equipment suppliers from *Voiding Equipment Guarantees/Warranties*, based on claims of poor power quality or acts of God: e.g. lightning.

If a power quality event occurs that exceeds "industry standards" (or exceeds limits that you define), our report will notify you of the event and describe the likely causes. This will allow you to take remedial action in a timely manner and possibly avoid damage to your equipment or facility, or prevent loss of critical data.

The report will include the date, time, and type of each power quality event. Depending upon the "level" of monitoring that you choose, it will usually indicate the direction of the event so that you will know if it occurred on the power company system or within your facility. This information could be particularly important if the event resulted in damage.

Figure 9-4 (2 of 2)

FIGURE 9-4

OUTLINE SUMMARY

- **Business announcement:** Make your business announcement outstanding. It should look like a wedding invitation on fine thick paper with raised letters.
- **Letterhead & business cards:** Create elegant letterhead and business cards. Parchment with raised letters and no logo would be a good choice.
- **Brochure:** Your brochure is the face of the company. It will evolve as your firm grows. Try to make it versatile and something that will make you proud and distinguish you from your competitors.
- **Mailing list:** Your mailing list is one of your most important promotional tools. Expand it to include those who may recommend you. Send all mailings to the entire list. Send a minimum of two per year, including updates, announcements, and holiday cards.
- **Lists & forms:** Create the lists and forms that your firm needs. Expand your contact list, prepare your government 330 form, and avoid construction headaches with your selected bidder list.
- **Promotional items:** Don't overlook personalized note paper, post cards, pens, tape measures, and anniversary stickers.
- **Articles:** Write and publish technical articles, send reprints to your mailing list, and include them with your brochure. Everyone will consider you to be an "expert" in your field.
- **Announcements:** Don't forget to send out announcements of awards, new services/specialties, and anniversaries.

CHAPTER 10

MARKETING METHODS

INTRODUCTION

You will discover both what to do and not do when marketing your new consulting engineering firm. Find out how easy it is to get leads and referrals. Understand which other potential clients are seeking your services, besides architects. See how developing a new "specialty" generates work for which you don't need to *compete.* Realize how public speaking opportunities are *free* publicity. Learn how entertaining clients and potential clients will garner loyalty. Ascertain why notes of congratulations and thanks and an occasional party will help your practice grow.

YOUR INITIAL PROJECTS

When you were creating your Business Plan, you did a Cash Inflow Projection (as described in Chapter 4 –Financial Forecasting). In estimating your income, you prepared a list of existing and potential clients and made some initial contacts to determine if you could count on their business. Have your notes from those meetings in hand when you contact them again. You need to start looking for work as soon as you have cast off (see Chapter 7).

YOUR WEBSITE

Much to my surprise, an engineering firm's website has become one of the most active sources of new clients. Some new clients will go to your website because of a referral or because they have heard that a competitor of theirs does business with your firm. However, you will find that many of your clients will find you while surfing the net.

CLIENTS

A client (obviously) is anyone who pays you for engineering services. I call clients who operate an "Architectural Practice" ARCHITECTURAL CLIENTS. I call all other clients PRIVATE CLIENTS (even though some of them may be graduate architects). They include hospitals, colleges/universities, federal agencies, housing authorities, trade contractors, attorneys/law firms, and many others.

SEARCHING FOR NEW CLIENTS

Hopefully, you are busy with new work, but now is the time to promote work from new sources. Expand your list of potential clients to include those whom you have not worked with in the past. First, look through the list of contacts that you have acquired since you started working, both in the course of business and in your work with engineering societies. Those whom you have impressed are often willing to introduce you to potential clients.

Continue attending some of the meetings of the engineering societies and industry organizations to which you belong.

Consulting engineers of a different "specialty" are often willing to share clients.

Land surveyors often find out about proposed projects even before an architect is selected. Making friends with them and periodically contacting them can provide good leads.

Search for leads on the internet and in newspapers.

Trade contractors are often asked by general contractors to bid on "design–build" projects, which include plans and specifications sealed by a professional engineer. Contact your trade contractors.

Housing authorities are great for upgrades (i.e., electrical, including fire alarm, generator, and lighting), but they usually hire architects for major renovations.

Hospitals are critical facilities and the work is specialized.

Higher educational facilities are recession-proof clients. The same comments as above apply here.

The FBO/CBD database (found at fbo.gov) is an online electronic database that can be searched for federal government architect and engineer opportunities.

If you are confident, knowledgeable, and well-spoken, serving as an expert witness should be part of your engineering practice. See Chapter 16 – Forensic Engineering.

OFFER SPECIALIZED SERVICES

What is a new specialty or new service? Well, it depends upon the type of engineering that you practice.

Our general specialty was the design of electrical systems for buildings. But we had many sub-specialties which we enumerated in our company brochure: lighting, electrical services, power distribution, fire alarm systems, emergency/stand-by power systems, uninterruptable power supply systems, etc. Also, if a potential client mentioned any kind of electrical system or building type, my response was always: "That's one of our specialties."

At various times in my career as an electrical consulting engineer, I offered the following new services: Electromagnetic Field Investigations & Mitigation, Power Quality Services, Uninterruptable Power Systems (UPS), and Arc Flash Calculations.

PUBLIC SPEAKING OPPORTUNITIES

Almost any kind of speaking engagement enhances your image and credibility.

Most professional organizations offer continuing education courses. It will enhance your reputation to arrange to be a guest lecturer.

Every time that you start a "new service," try to be a guest speaker at any professional organization whose members might use your service or recommend you.

Codes are not updated yearly; they are updated on a cycle, usually every three or four years. Consider holding a mini-seminar for your client's tradesmen and technical staff.

Major Seminar

When something ***dramatic*** happens, seize the opportunity.

In 2003, we had a tragedy in Rhode Island. 100 people perished in a fire at The Station nightclub. This catastrophic event was the impetus for a dramatic overhaul of our state fire code. The most revolutionary thing about this new code was that it was retroactive; the "grandfather clause" did not apply. All existing buildings were required to meet the new code.

I had recently been appointed to the "RI Building Code Standards Committee" and was too busy to take part in the overhaul of the fire code. But once it became law, I decided to put on a "free" seminar to educate fire officials and building owners of the new impact.

All my project managers were expert fire alarm system designers, and we had kept up with the proposed changes on a daily basis. But we spent a lot of time to make sure that we understood which building types required which upgrade and the required time table. I decided that my staff would be the "panel of presenters" and "field the questions." I personally introduced our services with a PowerPoint presentation, describing the advantages of hiring a consulting engineering firm to design fire alarm system upgrades.

I rented the largest banquet hall at the fanciest, centrally located hotel in the state for a grand buffet breakfast meeting. We did a mailing to our mailing list and to the mailing lists of all the Chambers of Commerce throughout the state. We did radio advertisements (1 minute spots liberally mentioning our company name) starting 10 days before the seminar. Note: Ethically, advertising by engineers is severely frowned upon, but these were "Public Service Announcements" sponsored by an engineering firm, so not a problem. I had many sleepless nights wondering if we would make incompetent fools out of ourselves and provide breakfast to only a handful of fire officials.

We had over 500 guests, and our seminar was an "enormous" success. It generated well over one half million dollars in fire alarm system design fees over the next 5 years and promoted other types of engineering work for us throughout New England.

I am certainly not saying to take advantage of the misfortunes of others. But I am saying to look for opportunities, have the courage to proceed, and do it *first class.*

Changing Attitudes

When my manuscript was reviewed by engineer friends, several told me that many potential clients resist forming friendships with their consultants. I had a few clients like that, and some hired my firm for all of their projects. Most of these were facilities directors who, I believe, were trying to avoid the appearance of undue influence regarding my continued selection.

However, I am being told that many of the selections are now being made solely based on "price." If you are in a market that seems to be unduly influenced by cost, make a concerted effort to standout from your competitors. Concentrate on the methods recommended in this book, including:

- Keep up-to-date with the latest technology.
- Gain credentials to build up your resume and brochure.
- Publish magazine articles, distribute reprints, and gain a reputation as an expert.
- Be the smartest guy in the room. Prepare white papers, memorize them, and share them with your staff.
- Prepare marketing materials, and implement methods that will make your firm standout from your competitors.
- Prepare an operations manual that emphasizes quality control and service to your clients.
- Hire carefully, train, and supervise your staff and treat them well and with respect. They are your greatest asset.

OUTLINE SUMMARY

- **Obtain E&O insurance**: Do no not "stamp" (seal) any documents until you have professional liability insurance (Errors & Omissions). See Chapter 8 – Legal, Accounting & Insurance.
- **Seek clients & projects:** First, recontact the potential clients that you visited when preparing your Business Plan.
- **Expand your sources for leads:** Include those on your contact list, consulting engineers of a different specialty, land surveyors, building inspectors, and utility company representatives. Also, search the internet and newspapers.
- **Seek private clients:** Include contractors of your specialty, housing authorities, hospitals, colleges and universities, property managers, manufacturers, developers, banks, law firms, and federal, state, and city agencies.
- **Offer new specialty services:** There is usually no competition for specialty services and no limitation on fees. Examples (related to electrical) include: EMF investigation and mitigation, power quality studies and monitoring, UPS systems, and arc flash calculations.
- **Always consider public speaking opportunities:** This is especially important when you are trying to promote a new service or new specialty. Also, hold seminars each time that the code of your discipline is updated. This shows clients that you are keeping up-to-date and gives you the opportunity to nurture your friendship with their tradesmen.
- **Seize an opportunity:** When something dramatic happens, be ready to take action. I am certainly not saying to take advantage of the misfortunes of others. But I am saying to look for opportunities, have the courage to proceed, and do it *first class*.

- **Entertain clients & potential clients:** If a problem occurs with the services of your firm, clients are more likely to tell you and allow you to make corrections if they know you and your spouse socially. They are also more likely to give the next project to a "friend."
- **Send notes:** Never forget to send thank you notes, and look for opportunities to send notes of congratulations. You don't even need to know someone to recognize their achievements.
- **Open houses and parties:** These are great ways to thank your clients, show off your celebrity clients, and tell all about new specialties and exciting projects.

CHAPTER 11

PROJECT PHASES

INTRODUCTION

Discover how to divide your contracts and your time cards into "phases" to simplify your accounting. Find out which phases are great for your "architectural" clients and those that work best for "private" clients. Realize when your specifications should include "or approved equal," "no substitution," "base bid," or "one of the following with no substitution." Learn how alternatives and your selected bidder list will avoid aggravation for you and your clients.

GENERAL

As stated previously, a client (obviously) is anyone who pays you for engineering services. I call clients who operate an "Architectural Practice" ARCHITECTURAL CLIENTS. I call all other clients PRIVATE CLIENTS (even though some of them may be graduate architects). They include hospitals, colleges/universities, federal agencies, housing authorities, trade contractors, attorneys/law firms, and many others.

Clients hire engineers to draw system plans for their buildings, including electrical, mechanical (heating, ventilating, air conditioning, and plumbing), fire protection (sprinklers), structural, and civil. Sometimes, very narrow specialties are required, such as acoustical. The electrical systems, for example, include utilities serving the building (power, telephone, data, cable TV, and fire alarm), lighting, power distribution, fire alarm systems, telephone distribution, cable TV distribution, and any other electrical system that the particular building might require.

The consulting engineer is responsible for designing within the client's budget limitations and coordinating with utility companies and inspection authorities. At the end of the design phase, the consultant prepares a specification document detailing the material requirements and system functions. During bidding, he or she attends pre-bid meetings, clarifies issues, and prepares addenda for the architect to issue to inform bidders of changes in the requirements. After a contract is awarded, the consultant reviews/approves shop drawings, detailing all equipment that his trade contractor proposes. During the construction phase, he or she visits the job site to record progress and clarify the contract documents. At the completion of the construction phase, he or she prepares a punch list detailing corrections to the work, if needed.

For "**architectural**" clients, I recommend the following breakdown:
(The percent indicates both the effort and the associated fee.)

MY BASIC METHOD OF ALLOCATING PHASES

- Study Phase – (separate optional phase)
- Design Phase – (75%)
- Bidding Phase – (5%)
- Shop Drawing Review Phase – (10%)
- Construction Meetings, Field Observations & Reports Phase – (10%)

For "**private**" clients, I used the following breakdown:
(The percent indicates both the effort and the portion of the associated fee.)

THE AIA METHOD OF ALLOCATING PHASES

- Study Phase – (separate optional phase)
- Preliminary Design Phase – (0–15%)
- Design Development Phase – (15–30%)
- Construction Documents Phase – (30–70%)
- Bidding Phase – (70–75%)
- Construction Administration Phase – (75–100%)

When you work for an "architectural" client, you are only responsible for a portion of the project. (In my case, electrical.) You work for the architect. He is in charge and writes the "front end" of the project manual, including the general conditions, bidding conditions, and the contract for construction.

Conversely, for "private clients," you are in charge and perform the functions of the architect. You write the "front end" of the project manual and coordinate the work of other consultants and, sometimes, the work of an architect, if needed.

I recommend the use of AIA documents. Therefore, it makes sense to break down the phases for private clients to match the AIA method.

OUTLINE SUMMARY

Architectural Clients

- **Phasing:** For "architectural clients," phasing includes study, design, bidding, shop drawing review, construction meetings, and field observations, including reports.
- **Study**: A study phase is more likely needed for renovations or additions. It is an evaluation of existing conditions to determine how the renovations and/or addition will be accommodated and often includes a written report and cost estimate.
- **Design:** The design phase includes detailed site investigation, design meetings (including conference memos), preparation of drawings, preparation of technical specifications, requests for approvals from utility companies and inspection authorities, and project coordination.
- **Bidding:** The bidding phase includes attendance at a pre-bid meeting, responding to questions from bidders, preparing addenda, and attendance at the bid opening.
- **Shop drawings:** The shop drawing review phase includes review and processing of shop drawings, responding to questions from contractors' staff, and preparation of clarification sketches or drawing revisions.
- **Construction meetings & field observations:** This phase includes attendance at job site meetings, field observations and reports, and a "punch list" final observation.

Private Clients

- **Phasing:** For "private clients," phasing includes study, preliminary design, design development, construction documents, bidding, and construction administration.
- **Study:** A study phase is a written report describing the proposed project, including options, a cost estimate, a time schedule (including design, bidding, and construction), and, usually, an 8 ½ x 11 sketch showing the relative locations of the major elements.
- **Preliminary design:** The preliminary design phase usually includes 15% progress prints, a preliminary design cost estimate, and a preliminary design project time schedule.
- **Design development:** The design development phase usually includes 30% progress prints, a design development cost estimate, and a design development project time schedule.
- **Construction documents:** The construction documents phase is the phase where you prepare the plans and specification for bidding and is the bulk of your work. The review meetings for this phase are similar to the other phases, and the number depend upon the scope and complexity of the project. Preparation of the specifications includes both the technical specifications and the front end of the project manual. Requests for approvals from utility companies and inspection authorities are also necessary as well as project coordination.
- **Restrictions on brands:** Review with the client what brands you plan to specify for major equipment. Also, review the specification method: "or approved equal," "no substitution," "base bid," or "one of the following with no substitution."
- **Alternatives:** Getting a price for adding or deleting certain items can avoid rebidding, if your estimate is not accurate.
- **Selected bidder list:** Avoid construction headaches by convincing your client to use your selected bidder list.

CHAPTER 12

FEE PROPOSALS

INTRODUCTION

You will realize why RFPs and RFQs are so important to both you and your clients and how to appropriately respond. Learn what questions to ask when your client is seeking a fee proposal. Understand which fee methods are customary, which methods to avoid, and why an estimate of hours is essential. Comprehend how simple it is to uncover the architect's compensation and the fee that he expects to pay you. Finally, discover the existence of never before published "Difficulty Tables" and "Fee % Curves," developed over 35 years of professional practice.

RFPS & RFQS

RFPs are Requests for Proposals and RFQs are Request for Qualifications. In both cases these are notices of the need for architects and/or engineers. Both types are advertised in newspapers, or in the FedBizOpps (FBO) database, or elsewhere. Both require a submission of qualifications. But, an RFP usually also requires a submission of a fee proposal or an hourly rate schedule.

You may become aware of these by calls from previous clients, potential new clients, or from engineers of a different specialty.

PROJECT SCOPE

When you review the proposed project with the client, take careful written notes. If the client doesn't mention the following in his explanation of the project, ask:

- Is this a single-use building or multi-use?
- What is the area (sq. ft.) of each use?
- What is the area (sq. ft.) of "new," "renovation," and "addition"?
- What is the project budget?
- What is the project schedule?
- When will we be able to get started?
- What is the schedule for each phase: study, preliminary, design development, construction documents, bidding, and construction administration?
- How long is construction expected to last?

FEE AGREEMENT METHODS

Often, the client dictates the method of payment. The following are the most common options:

Fixed-Fee

This is a specific dollar amount for each phase of the project. I found this to be the most common. Assuming that you carefully define the scope and detail assumptions and exceptions, it avoids arguments and gives the client a specific fee amount. Occasionally, there is a dispute about the percent complete of each phase for monthly billing. But, assuming the project doesn't get canceled, that problem eventually gets worked out.

Percent of Construction Cost

Your fee is a percent of the "low bid" for the construction of your discipline/ specialty. Some of my clients used this method. It is important to state that the construction cost includes all "designed alternatives." For example, let's assume that a decision is made to get an "additive alternative price" for an emergency/ standby generator and emergency distribution system in lieu of battery units. The electrical engineer should get a design fee based on a percent of the cost of both designed options.

Hourly – (not-to-exceed)

Avoid this whenever possible; it always causes arguments. The not-to-exceed number is never more than what a fair fixed-fee would be. Therefore, there is no "upside" for you. If you are honest, you will soon go out of business.

Others

Others are "per diem," "multiple of payroll cost," "hourly without limit," etc. State DOT (Department of Transportation) projects are often done on a multiple of the payroll method. But, periodically, they audit your books and often disallow expenses (like rent for your office space in a building owned by your company or that you own personally). When this happens, get out your check book. Consider methods of payment carefully before accepting assignments. Never accept a project on a "handshake" or a promise that "we will work it out later."

ESTIMATE HOURS

Regardless of your fee payment method, ALWAYS estimate the hours of work that the project is expected to require.

VERIFY CONSTRUCTION COST

You will need up-to-date ESTIMATING BOOK(S) or CDs that include:

- General New Construction Cost ($/sq. ft.).
- Costs for your trades (percent of total or $/sq. ft.).
- A geographical multiplier.

RS Means Company www.rsmeans.com is a company that produces estimating books and CDs that are updated annually.

ESTIMATE THE ARCHITECT'S FEE & YOUR ASSOCIATED FEE

The first step is to estimate the "difficulty" of this type of project for the architect. This is not an exact science, but I have prepared Figure **12-1** "DIFFICULTY LIST" (Included in the Complete Guide) to assist you.

The second step is to estimate the "general" construction cost of each of the areas of the project. Use the latest version of one of the many good "Construction Cost Estimating Books." (Unless you can find a "combined book," you will need two or more books: one for general construction and others for your trades.) They each have two related tables: one is the average sq. ft. cost and the other is an adjustment factor for your geographical area. Be sure to use both.

The next step is to enter the difficulty and the construction cost on the correct FEE PERCENTAGE GRAPH.

See Figure **12-2** "FEE % GRAPH FOR NEW CONSTRUCTION." (Included in the Complete Guide)

See Figure **12-3** "FEE % GRAPH FOR RENOVATIONS." (Included in the Complete Guide)

Example

This may sound a little complicated, but an example will reveal how simple (but not exact) it really is.

Assume that the project is an existing high school located in the state of Rhode Island. The new project is to renovate a portion of the existing school and add a new library.

- Existing building size: 40,000 sq. ft.
- Portion to be renovated: 10,000 sq. ft.
- Library addition: 15,000 sq. ft.

First, convert the above areas to "general" construction cost.

- High school: $197/sq. ft. x 1.22 x 10,000 sq. ft. = $2,403,400
  (Note: I used the factor 1.22 to adjust the costs for the state of Rhode Island. You would, of course, adjust for the geographical area of your project.)
- Library: $215/sq. ft. x 1.22 x 15,000 sq. ft. = $3,934,500
  (Note: I have ignored the 30,000 sq. ft. of existing high school that is not being renovated. As a consultant, you will have to spend some time assessing existing systems and possibly designing replacements or modifications if these systems are not expandable to meet the needs of the new project. But the architect will probably not get more fee since his/her work in these areas is minimal. However, include your extra time in your budget of hours.)

Second, use Figure **12-1** to determine the "difficulties."

- High school = "C"
- Library = "D"

Next, use Figure **12-2** to determine the architect's fee for the library addition because this is new construction.

- Library addition – 7.5%

Use Figure **12-3** to determine the architect's fee for the renovation.

- High school renovation – 9.8%

Then, determine the budgets for mechanical and electrical construction.

- Library addition
    - Mechanical: 15% x $3,934,500 = $590,175
    - Electrical: 9% x $3,934,500 = $354,105
- High school renovation
    - Mechanical: 18.1% x $2,403,400 = $435,015
    - Electrical: 12.9% x $2,403,400 = $310,038

The architect (or other prime professional) expects you to work for a "**portion**" of his fee. When I first started in practice, it was ⅔ and later increased to ¾. You need to check with other consultants in your area of practice to determine what the architect is expecting. I have used ¾ in this example.

Finally, the mechanical and electrical engineering fees are determined.

- Mechanical:
  Library addition: 7.5% x ¾ x $590,175 $33,197
  HS renovation: 9.8% x ¾ x $435,015 $31,973
  Total mechanical fee $65,170
- Electrical:
  Library addition: 7.5% x ¾ x $354,105 $19,918
  HS renovation: 9.8% x ¾ x $310,038 $22,787
  Total electrical fee $42,705

These are the "approximate" fees that the architect should be expecting from his mechanical and electrical consultants. Compare these to your "estimate of hours." If they differ substantially, recheck both.

Most firms are concerned about paying hourly employees "overtime." I never was. 5–10 hours of overtime per week makes most employees happy. But any more than that makes them less productive. However, I believe that limited overtime actually improves your bottom line. Your employee's hourly billing rate should be 2.5 to 3 times what you actually pay them in order to cover benefits, taxes, overhead, and profit. If you pay an employee $60/hour, his billing rate is about $175/hour. That rate covers all the items listed based on a 40 hour work week. If he works overtime, you pay him $60 x 1.5 = $90/hour, plus a little extra in payroll taxes, electricity, and heat or AC. But the cost to you, for that extra hour, is a lot less than $175.

Some projects are "losers." Try to pass up most of them.

I have just shared with you the fee "secrets" that took me 35 years to perfect. Apply my advice and use your best judgment; your proposals will win you work.

OUTLINE SUMMARY

- **RFPs & RFQs:** Be on the "look out" for RFPs and RFQs and respond ASAP.
- **Sub-consultant:** For requests where you will be the "sub-consultant," notify your appropriate clients and potential clients who may want to be the "prime consultant." Express an interest in being on their team along with the documentation showing why your firm is the best choice as consultants.
- **Prime professional:** When you are qualified to be the "prime professional," respond with your RFP-RFQ Response Letter and attachments.
- **Fee proposal scope:** When clients ask for a fee proposal, bring a list of questions to the meeting. It will assist you in gathering the necessary information.
- **Fee methods:** Most fees are a fixed amount or a percent of construction; avoid other methods.
- **Estimate hours:** For <u>every</u> project, estimate hours, regardless of the method of fee payment. Develop your own form for the purpose.
- **Architect's fee:** Estimate the architect's fee based on the "Difficulty Tables" and "% Curves" that were developed over 35 years of professional practice.
- **Your fee:** Estimate your fee based on a portion of the architect's fee and the calculated construction costs of your trades.
- **Fee proposal:** Finally, prepare your "winning" fee proposal letter.

CHAPTER 13

OPERATIONS MANUAL

INTRODUCTION

Understand how your Operations Manual is the "guidebook" to your engineering practice and will be a major ingredient in your firm's success. Learn how to organize yours.

WHAT IS AN OPERATIONS MANUAL?

Your Operations Manual contains the forms, procedures, issues, letters, calculations, questionnaires, and policies of your firm. It will be unique to your practice and, if thoughtfully prepared and well organized, will be a major ingredient in your firm's success.

The following is the table of contents that I used. Customize it to suit your needs.

OPERATIONS MANUAL

Table of Contents

OUTLINE SUMMARY

- **Operations manual:** Your Operations Manual is the "guidebook" to your engineering practice and will be a major ingredient in your firm's success.
- **Office forms & procedures:** When training new employees to the basics of your operation, have them review the office forms and procedures.
- **Technical procedures:** Organize the technical step-by-step approach to managing each assignment under technical procedures.
- **Codes & technical issues:** Most code requirements are clear, but some are subject to interpretation, and it is important that a firm be consistent. Carefully documenting code research and include it in this section to save time on future projects.
- **Approval letters:** Review your projects with utility companies and inspection authorities, and carefully document this process.
- **Coordination sheets**: Create a series of forms to coordinate major systems and equipment with the other project consultants, vendors, and the architect.
- **Forms & form letters:** Customize forms and form letters for your disciplines to make your firm more consistent and efficient.
- **Calculations:** Document your calculations with forms tailored for your practice.
- **Questionnaires:** Make organized lists of questions to ask at meetings or during conversations with utility companies and clients.
- **Private clients:** The forms and other materials that will assist you in this increased responsibility as the "prime professional" are kept in this section. Partially fill out these forms with information that is standard for your projects.
- **Office policy:** Explain to your staff the general policies of your office in this section.

CHAPTER 14

DRAWINGS, SPECIFICATIONS & PROJECT MANUALS

INTRODUCTION

You will discover how logical sheet organization will please the contractors and get more of them to bid. Learn how standard details save time and increase profits. Discover how adding addenda and change orders to your record drawings and project manual will avoid costly oversights. Find out how developing a "specifications questions list" will make specification writing easier and avoid errors. Ascertain how to create "three" specification versions to serve all of your needs. Observe how to acquire a "front end specification" for your project manual and how to easily customize it for each project.

GENERAL

The requirements for construction projects are set forth in three documents. This concept is clearly defined in Document A201™ General Conditions of the Contract for Construction by The American Institute of Architects, as quoted herein:

> § 1.1.5 **The Drawings**
> The drawings are the graphic and pictorial portions of the Contract Documents showing the design, location and dimensions of the Work, generally including plans, elevations, sections, details, schedules and diagrams.

§ 1.1.6 **The Specifications**
The specifications are the portion of the Contract Documents consisting of the written requirements for materials, equipment, systems, standards and workmanship for the Work, and performance of related services.

§ 1.1.7 **The Project Manual**
The Project Manual is a volume assembled for the Work which may include the bidding requirements, sample forms, Conditions of the Contract and Specifications.

DRAWINGS

As stated in Chapter 11 – Project Phases, if you are working for an architect, your "engineering drawings" get started when the architect sends the CAD/BIM files of the building floor plans (shells) to you. When you are the prime professional, you start the drawings yourself.

When I first started my practice, we had to trace these shells on a light table to make our own "originals" (three times: once for power, then for low voltage, and again for lighting). We also had plenty of erasers handy for the inevitable changes. We thought we were getting high tech when we got an "electric eraser." In addition, each engineering discipline has additional sheets: legend, symbols, schedules, notes, 1-line diagrams, etc. Again, in my early days, these were all custom drawn by hand. Later, we used "stick-ons" (stand-pats). More and more information keeps getting added (and changed) until the drawings are finally complete.

Now, of course, we have CAD/BIM and fast computers to make the process more flexible and changeable. Each discipline/specialty has its own unique design and presentation methods and needs, which are beyond the scope of this book. But there are some commonalities to all the design specialties.

SPECIFICATIONS

As stated in Chapter 11 – Project Phases, the technical specifications are a written document that details the requirements of each of the items shown on the drawings. It specifies the requirements of something as (seemingly) simple as a light switch to something as complex as the functions of a fire alarm system. It also enumerates requirements for manufacturer's experience, workmanship, testing, and commissioning. This document is added at the end of the project manual by your architect client, unless you are the lead/prime professional, in which case the whole project manual is your responsibility.

I eventually had three (3) specifications versions:

- My standard one-section specification
- A modified multi-section MasterSpec Professional specification
- A modified multi-section MasterSpec Outline specification

PROJECT MANUAL

For "private client projects" you will need to produce the entire project manual because you are the lead/prime professional. Therefore, you will need both bidding requirements and contract documents. The most commonly used are the ones produced by the AIA or the NSPE.

I immediately thought, "I'm an engineer; I should use NSPE standards." Fortunately, I asked around, and the president of one of the larger local engineering firms told me that when they started using the NSPE method, they got so many complaints from contractors that they changed to the AIA method. Basically, the contractors were already familiar with the AIA bidding conditions and contract documents and didn't want to learn another system. Also, many institutions (universities, hospitals, and municipalities) prefer the AIA documents.

Therefore, I used the AIA System.

OUTLINE SUMMARY

- **Common presentation:** Adopt a "common drawing presentation" for all of your projects to make it easier for bidders to find things.
- **Standard details:** Implement "standard drawing details" to make your office more efficient.
- **Addenda & change orders:** Update your "office record drawings and record specifications" to include addenda and change orders.
- **Specifications questions:** Produce a list of "specification questions" to make specification writing easier and more accurate.
- **Three specifications:** Develop three specification versions: one-section, multi-section, and outline.
- **General requirements:** Add a general (specialty) requirements section to your specifications. Locate this near the beginning of your "one-section" specification and as a separate section in your "multi-section" specification.
- **Project manual:** Acquire a project manual for private client projects, including bidding requirements and contract documents.
- **Insert list:** Create a "Project Manual Insert List" to simplify preparing the project manual and the accompanying cover letter to the client.

CHAPTER 15

HUMAN RESOURCES

INTRODUCTION

Discover both the benefits and drawbacks of having employees and why you should grow your firm. Find out how to conduct an effective employment interview. Detect the art of making your offer "confidential." And last, but not least, uncover how to prevent a firing from backfiring.

The people who work for you are referred to as "human resources"; I just called them my staff. It is kind of like a marriage: "You can't live with them, but you can't live without them, either."

WHY HAVE HELP?

Benefits

- You are able to avoid "mundane" tasks (deliveries and fetching).
- You can assign tasks that you don't like to do.
- Clients object to paying your hourly rate for simple tasks.
- The lasting value of a "service" business is in the staff. *A mechanical engineer friend started his practice about the same time that I started mine. He made a good living, but he never had more than one drafter. When he retired, he laid off the drafter and closed the office. By the time that I retired, we had grown to a staff of eleven, and I had something valuable to sell.*

Drawbacks

- They don't get along with each other or, sometimes, with your clients.
- They often become your biggest deterrent to doing business.
- Dealing with their problems will consume more time than technical issues.
- As mentioned in Chapter 1, before you know it, you are responsible for thirty or more mouths to feed. That is an awesome and burdensome responsibility.

INTERVIEWS

I called potential employees personally to arrange for an interview. I tried to get their home phone number (or cell phone number) and called them after working hours.

When they arrived for the interview, my office manager sat them in the conference room to fill out our employment application. When the application was complete, we went into my office for the interview. After reviewing the application, I had a typed list of questions. Just because the applicant's title is "project manager," don't assume the applicant does all the things that your project managers do. Question specifics. How much of the design do you do on your own? Do you attend meetings with the client? If yes, do you go alone? Do you write letters, conference memos, specifications, etc.? Also, ask how long the applicant has been doing each of these tasks. Take careful notes.

Tell the applicant that it is your policy to do a credit and background check on all potential employees. Ask if anything unusual will show up that they would like to explain. Make sure that you have a place on your employment application for them to sign authorizing both. I had a few people refuse to authorize this and they left before the interview. Great! This saves time and avoids a hiring mistake. With one background check, I received a copy of the applicant's "mug shot."

THE OFFER

Once that you have decided to make an offer, call the candidate at home. Say that you are impressed and want to make an offer. Then, say, "But first, I want to tell you what I tell all prospective employees. You are, of course, free to discuss this offer with your significant other. But I want you to agree not to discuss it with your present employer or anyone else until you give me a definite answer. Is that acceptable to you?" If they don't agree, don't make an offer. If they do agree, repeat the condition one more time before making the offer. However, don't be shocked if they call you back and say, "My present employer has agreed to match your offer, and I have decided to stay."

FIRING IS HARDER THAN HIRING

The construction industry is "cyclical." It fluctuates with the economy.

When we had to lay someone off, I had a senior associate tell me (more than once) that an employee hadn't really worked out; after two years, he still worked at half the speed of the others. When I asked, "Why did we keep him this long?" he said, "Well, he was better than no one." On a very short-term basis, to meet a deadline, that might make sense, but to have this persist for two years defies logic. Don't assume that your #2 guy is going to tell you when you have some "dead wood." Conversely, make better hiring decisions than I did.

Whenever I let someone go, I tried to do it at the end of a work day. I had an "exit interview" form, and I usually said the reason was that "several projects that were about to start have been canceled." I paid them to the end of the week plus two weeks' severance pay. In addition, I offered them an additional two (or more) weeks' pay in return for signing a waiver agreeing that they had no grievance with the company and would not sue me or the company for any reason. If the departing employee was a project manager, I gave him a "project status" form for each of his projects and asked him to review his job folders and list any pending items. It was never fun, but I always did it personally. It was really difficult when they cried. But it was worse when they thanked me for the opportunity that I had given them.

OUTLINE SUMMARY

- **Benefits:** The benefits of having employees include avoiding mundane tasks and having clients objecting to paying your hourly rate for simple responsibilities. Also, the lasting value of a "service" business is in the staff.
- **Drawbacks:** The drawbacks of having employees include them not getting along with each other or, sometimes, with the clients, dealing with their problems, and being responsible for thirty or more mouths to feed—an awesome and burdensome responsibility.
- **Small size:** The benefits of being small include knowing what your staff are doing, seeing how they are developing, and being aware of how each project is progressing. The drawbacks of being small include too many questions/interruptions and your staff overhearing your conversations.
- **Growth:** The benefits of growth include privacy and increased opportunities. The drawbacks of growth include becoming less aware of what is going on.
- **Potential employees:** Recommendations for new employees come from salesmen, your existing employees, and contractors on your "Selected Bidder List." In addition, consider hiring tradesmen.
- **Interviews:** Invite potential employees to an interview when they are at home, develop an employment application and question list, and require a credit and background check.
- **The offer:** Try to require confidentiality when making an offer.
- **Company night:** Occasionally, socialize with your staff as a group outside the office; foster a "family" atmosphere and they will be loyal and want to remain part of the clan.
- **Firing is the hardest task:** Conduct an organized exit interview, pay severance, offer extra payment in return for signing a release, and have them fill out project status reports.

CHAPTER 16

FORENSIC ENGINEERING

INTRODUCTION

Understand what "forensic engineering" is and discover why this exciting, challenging, and profitable field should be part of your consulting engineering practice. Find out what is involved and if you have what it takes to be a success in the litigation arena.

You will learn how to prepare a curriculum vitae (CV), formulate a forensic fee agreement, conduct investigations, write reports, and prepare for depositions and the trial, including cross-examination.

WHAT IS FORENSIC ENGINEERING?

Forensic engineering is defined by the National Academy of Forensic Engineers (NAFE) as "the application of the art and science of engineering in matters which are in, or may possibly relate to, the jurisprudence system, inclusive of alternative dispute resolution." These engineers serve as consultants to the legal profession and as expert witnesses in courts of law.

SHOULD YOU CONSIDER THIS FIELD?

Benefits

- Interesting work
- Low liability
- High hourly rates
- Advanced payment via retainer
- Little competition

Drawbacks

- Can be stressful
- Sometimes requires travel

Because litigation consulting is interesting, challenging, and profitable, I recommend adding "forensic engineering" to your practice.

DO YOU HAVE WHAT IT TAKES?

To be an "expert witness," you don't need to be an engineer, but you do need experience and credentials in your field. To be an "engineering expert, you will usually need an engineering degree and a license as a Professional Engineer. In addition, being a member of NAFE is a big help in being selected and in assuring acceptance as an expert in court.

It is also important to have knowledge and qualifications in the engineering specialty involved in the particular case at hand. For example, if the issue is lighting or illumination, experience in lighting design and membership in the Illuminating Engineering Society of North America (IES) would be helpful. It's best to stick to your general area of expertise. If you are an electrical engineer, for instance, don't take on a civil engineering case.

In addition, you should possess the following general traits:

Speaking/teaching ability

In court you need to speak clearly, using proper English and making your statements and replies concise and easy to understand. You will be teaching the judge and jury your opinion of the case and it is your job to convince them that you are correct.

Writing skills

Written reports need to be clear and professional. Don't put anything in writing until your attorney-client requests. Be sure you can defend every word under cross-examination.

Willingness to prepare

You are being paid to study the case and the technical issues involved. This can often be tedious, but it is the key to success. When preparing for court or deposition, it is crucial that you prepare for all possible questions and memorize much of the information.

Reading skills

In most cases you will receive a box full of documents to review. These may include: (1) affidavits (written declaration of facts sworn to); (2) complaint (plaintiff's initial pleadings); (3) depositions (oral testimony taken by the opposing attorney in advance of trial); (4) indictments (written accusations presented by a grand jury); (5) interrogatories (written questions sent to the opposing side and written answers submitted under oath); (6) petitions (written applications to the court requesting judicial action); (7) pleadings (written statements of contentions of the parties in the suit); (8) subpoenas (written orders for witnesses to appear); (8) summons (writ directing an officer to notify a defendant to appear in court); and (9) transcripts (official record of proceedings in a trial, deposition, or hearing).

It is part of your job to read all of these documents and to cull the information pertinent to your involvement in the case. This can mean many hours of reading even though many of these documents have no bearing on your portion of the case.

WHAT IS INVOLVED?

Cases include some, or all, of the following: site investigation, examination and testing of evidence, reports (verbal, summary, and detailed), testimony at deposition, and testimony at trial (direct and cross-examination).

There are two sides to every case. The person who brings the action is the "plaintiff." If you work for his or her attorney, you are a plaintiff's witness. The other party in the suit is the "defendant." If you work for his or her attorney, you are a defendant's witness. In your forensic practice, you will almost exclusively be involved with civil cases (noncriminal matters).

There are two types of witnesses: a "fact witness" and an "expert witness." A fact witness can only tell what he or she observed, not what others have said (hearsay) and not his or her opinion concerning the case. You, however, will be an expert witness (a technical witness), and, as such, you are allowed to express your opinions on matters within your expertise.

Above all, you must be truthful at all times. You must be supportive of your opinion without arguing for your attorney/client. Your opinion will not always coincide with that of your attorney/client. His or her job is to advocate for his client. Your job is to form an accurate and scientifically consistent opinion and to convey that to your attorney/client. If your attorney/client doesn't agree with your theory of the case, he or she can get another expert. In addition, he or she can benefit from your contrary opinion because your findings will point out the weaknesses in their case and may prompt a settlement.

CURRICULUM VITAE (CV)

Your CV is what your expert witness resume is called. The more detailed and complete it is, the more qualified you will appear.

Include the following items:

- Specialty – Summary of your area of expertise.
- Education – Starting with college and including degrees and honors, post-graduate studies, and certifications.
- Professional registrations – List states and numbers.
- Professional practice – Present and previous company affiliations, positions, and titles.
- Professional affiliations – Memberships in professional/technical groups and offices held.
- Honors and awards – List name of honor/award and group presenting it.
- Publications – List title, publication, and month and year.
- Lectures – List title, group presented to, and month and year.
- Forensic activities – List the most recent first.

See Figure **16-1** for the first few pages of my CV.

CV EXAMPLE (1 OF 4)

(Example of Forensic CV)

JOHN D. GASKELL, P.E.

Specialty: ELECTRICAL SYSTEMS FOR BUILDINGS

(Forensic consulting electrical engineer specializing in disputes relating to Electrical design and construction of buildings: Electrical failures/explosions; electrocutions; fires of electrical origin; errors and omissions; standard of care; National Electrical Code compliance; illumination evaluations; power quality; grounding; lightning protection; and electrical accidents. Member: National Academy of Forensic Engineers.)

Education: Bachelor of Science in Electrical Engineering
University of Rhode Island, 1966

Associates in Electrical Engineering Technology
Wentworth Institute, Boston, MA 1962

Professional Registration:

Registered Professional Engineer in Massachusetts and Connecticut.

Professional Practice:

John D. Gaskell, P.E. is president of Gaskell Associates Consulting Electrical Engineers. This engineering practice, established by him in 1971, specializes in electrical systems for buildings. Gaskell Associates is a division of Thielsch Engineering, a 250 person, diverse engineering firm.

For over 30 years Gaskell Associates has provided electrical design services to architects and building owners, including: Power Distribution Systems; Emergency Power systems; High voltage Power Distribution systems; Lighting Design; Fire alarm systems; Lightning Protection Systems; Security alarm Systems; communications systems; MATV & CCTV Systems; Computer Power and Network Systems; UPS & Power Conditioning Systems; Magnetic Field Measurements and Mitigation; Power Quality Studies, Measurements and Monitoring; Feasibility Studies and Forensic Investigations & Legal Testimony.

Service to clients has been the key to their success.

FIGURE 16-1

CV EXAMPLE (2 OF 4)

(Example of Forensic CV)

Professional Affiliations:

National Society of Professional Engineers (NSPE)

President of the Rhode Island Society of Professional Engineers 1980-1981

Illuminating Engineering Society of North America (IESNA)

Founding President of the Rhode Island Chapter 1975-1976

National Academy of Forensic Engineers

R.I. State Building Code Standards Committee (Past Member)

Electrical Sub-committee Past Chairman

Electrical League of Rhode Island

Providence Engineering Society

National Electromagnetic Field Testing Association

Honors and Awards:

FREEMAN AWARD from the Providence Engineering Society

This award was established for the purpose of recognizing major achievement in engineering. It is held to be unique in that it recognizes accomplishment, inventiveness, development of theory, or the application of scientific knowledge in such a way as to provide advancement in the art and practice of engineering.

ENGINEER OF THE YEAR

From the Rhode Island Society of Professional Engineers

LIGHTING DESIGN AWARDS

Hall Library, Cranston, Rhode Island
From the New England Section of IESNA

Lincoln School Gym, Providence, Rhode Island
From the Rhode Island Chapter of IESNA

ESD FIRST PLACE NATIONAL DESIGN AWARD

"UNINTERRUPTIBLE Power Supply" (UPS) system for Fleet National Bank's Computer Operations Center in Providence, RI, with initial capacity of 600 KVA and expandable to 1 Megawatt.
From Electrical Systems Design Magazine

FIGURE 16-1

CV EXAMPLE (3 OF 4)

(Example of Forensic CV))

WEISMAN MEMORIAL AWARD

In recognition of sustained and meritorious service
From the Rhode Island chapter of IESNA

MAN OF THE YEAR

From the Electrical League of Rhode Island

Publications:

"Anatomy of a Power Outage." Electrical Construction & Maintenance (EC&M) Magazine, October 2001

"Consulting Engineer: Watch Out for Decline in Power Quality." APC UPTIME, April 2001

"Avoiding EMF's – Designing Buildings Limiting Occupants Exposure to EMF's." Power Quality Assurance, September/October 1996

"Electrical Design with EMF in Mind." Electrical Construction & Maintenance (EC & M) Magazine, February 1995

"Expandable UPS Serves Bank Operations Center." Electrical Construction & Maintenance (EC & M) Magazine, March 1988

"Effects of Voltage Reduction." Electrical Consultant, February 1974

Lectures:

"How Fire Alarm System Requirements Affect Building Owners." Lecture presented to the Building Owners and Managers Association of Rhode Island, May 2005.

"What Realtors need to know about the New Fire alarm Systems Requirements." Lecture presented to the Rhode Island Association of Realtors, February 2005.

"Fire Alarm Systems for Building Owners." Seminar presented by Gaskell Associates Consulting Engineers to over 400 attendees regarding the changes to the Rhode Island Fire Safety Code precipitated by the 2003 Station Nightclub Fire, November 2004.

"Electrical Systems for Buildings." Lecture presented at the advanced lighting course for the Rhode Island Chapter of the Illuminating Engineering Society of North America, January 2000.

"Power Quality Audits and Monitoring." Lecture presented at the Rhode Island Plant Engineering Maintenance Show and Conference, October 1999.

"Forensic Engineering – How to prepare a case." Lecture presented at the Rhode Island Plant Engineering Maintenance Show and Conference, October 1999.

"Power Quality Studies and Monitoring." Lecture presented to the Southern New England Network Users Group, July 1998.

FIGURE 16-1

CV EXAMPLE (4 OF 4)

(Example of Forensic CV)

"Power Quality in New Buildings." Lecture presented at the Building Systems Course, Roger Williams University, Bristol, RI, April 1998.

"Electrical Systems for Buildings." Lecture presented at the advanced lighting course for the Rhode Island Chapter of the Illuminating Engineering Society of North America, January 1998.

"Power Quality – Why be concerned." Lecture presented to AIA Rhode Island, a chapter of the American Institute of Architects, November 1997.

"Forensic Engineering." Lecture presented to the Rhode Island Chapter of the Illuminating Engineering Society of North America, November 1997.

"Power Quality Audits." Lecture presented to the New England Healthcare Engineers Society at their Fall Seminar in New London, Connecticut, November 1996.

"The Industrial Design Building @ 161 South Main Street." Lecture at the Rhode Island School of Design (RISD), October 1996.

"Power Quality Studies & Solutions." Lecture presented to the Rhode Island Chapter of the American Association of Plant Engineers, October 1996.

"Avoiding EMF's." Lecture presented at the National Conference on Harmonics and Power Quality in Philadelphia, Pennsylvania, May 1995.

"The Architect's Role in Electrical Design." Lecture presented at the Rhode Island School of Design (RISD), October 1994 and December 1995.

"Lighting Design in Commercial Buildings." Lecture presented at Roger Williams University College of Architecture, September 1992.

"I.E.S. Lighting Design Awards." Lecture presented to the Rhode Island Chapter of the American Institute of Architects at their Fall Conference, October 1990.

"Uninterruptible Power Supply Systems." Lecture presented to the Rhode Island Hospital Engineers Association, February 1987.

"Energy Efficiency in Electrical Systems." Lecture presented for the Rhode Island Chapter of the American Association of Plant Engineers, March 1986.

"Electrical Systems for Buildings." Course presented at Roger Williams University College of Engineering, fall 1982, 1983 and 1984.

"IES Introduction to Light and Lighting." Course for the Rhode Island Chapter of IES, Spring 1978.

FIGURE 16-1

FORENSIC FEE AGREEMENT

See Figure **16-2** for an example of a Forensic Fee Agreement. Most of these items are self-explanatory. Modify, as needed.

Instead of using my fee agreement, you may want to consider the SEAK Expert Witness Retention Contract. I have been told that a large number of experts use it and that it is well accepted by attorneys, despite offering fair and favorable terms for the expert. It is available for download on the website http://www.seak.com, for a fee.

FORENSIC FEE AGREEMENT

(Your company Letterhead)

CASE: ______________________________

SCOPE OF SERVICES: Electrical Consultant and (if desired) as Expert

SERVICES & EXPENSES

Personnel	Rate per Hour
Forensic Engineer	$XXX
Forensic Investigator (2)	$XXX
Technician	$XXX

Expenses	Rate
Travel (3)	At cost
Mileage	$0.XX per mile
Out of Pocket Expenses	At cost

(1) Rates indicated are for all services, including travel. A minimum of eight hours per day is charged for attendance/waiting/testimony at any deposition, trial or hearing, regardless of location, and regardless of any delays or rescheduling, and shall not be contingent on actual testimony or testimonial qualifications. For depositions we will provide you with a separate statement with which to bill the deposing attorney for re-imbursement.

(2) A Forensic Investigator is a non-P.E. who will assist the chief investigator (expert witness) with investigations.

(3) Travel expenses include car rental, taxi, meals, business class hotel and coach class air travel. Travel arrangements and lodging selections are subject to approval by us.

TERMS

- Services are subject to availability.
- To retain us, please send $X,XXX to establish a Retainer fund, which sum, exclusive of any expenses, shall constitute our minimum professional fee for availability to you in this matter. Our hourly fee, billed in one-half segments, and any expenses shall be drawn from the Retainer fund as work is completed.
- Periodic detailed billing statements are rendered, usually monthly, and are due immediately upon receipt. Outstanding balances past due over thirty days are subject to a delinquency charge of xx percent per month until paid. (Insert your company name here), with liability, may withhold delivery of reports and other data and may suspend performance of its obligations to Client pending full payment of all charges.
- In order to schedule our deposition or trial/hearing testimony (regardless of location), or any out of state travel, all payments must be current and advance payment must be received for all expenses plus fees for time anticipated.
- The laws governing this contract shall be those of the State of (insert your state) and the venue and jurisdiction of this contract is the county of (insert your county). You agree to be responsible for all attorney fees, costs, expenses, and interest relating to the collection and enforcement of this contract.
- In retaining us, not withstanding any relationship you may have with third parties, contingency arrangements, subrogation, etc., you, as our client, agree to be liable for and to pay us for any and all services rendered and expenses incurred, including any fees or costs incurred by our response to discovery efforts of others. Terms on outstanding balances are net 30 days.
- The client (and subsidiaries of the client) agree not to attempt to hire (or discuss employment with) any employees (or former employees) of (insert your company name), until one year has passed, since their employment by us.
- We guarantee that we will use our best efforts on your behalf; however, we do not promise or guarantee what our eventual opinions will be, nor what effect our work will have, on the outcome of the case. Our conclusions will be based, in part, on the accuracy and completeness of the information you provide.

These fees and provisions are hereby agreed to by the undersigned parties.

____
Insert Your Company Name

Authorized Signature: ______________________________

Printed Name & Title: ______________________________
Address: ______________________________

Dated: ______________________________

Dated: ______________________________

FIGURE 16-2

INVESTIGATIONS

The more investigation, research, analysis, and testing that you do, the more credible and convincing your testimony will be. Your investigation will, of course, depend on the details of the issues in your field of involvement. Your attorney/client can arrange access to evidence or the site. Do not rely on your memory; take detailed notes.

REPORTS

Do not write reports until your attorney/client requests. There are three basic types of reports: verbal, summary, and detailed.

THE DEPOSITION

Depositions (oral testimony, under oath, taken by the opposing attorney in advance of trial) are often required of expert witnesses, especially if the case is expected to go to trial.

THE TRIAL

The good news is that the vast majority of cases settle; the bad news is that you need to prepare every case as if it will go to trial.

Review what to bring to court with your attorney/client. Usually, bring the following:

- Your CV. Be prepared to present it orally from memory. Summarize your cases by saying something like, "I have been involved with 18 cases as an expert witness. I was deposed in 10 of them, and I testified at trial 4 times."
- Your fee schedule, time sheets, bills, and case summary. Make sure that these records are up to date and be prepared to answer the question, "How much time did you spend on this case and what were your total charges to date?"
- Your report. Again, be prepared to present it orally from memory.

Don't bring your case file unless directed to do so. If you do bring it, carefully remove any irrelevant, early, or incorrect information.

Practice aloud in front of a mirror until you can recite the answers from memory with conviction and without "umms" and "I thinks." This will adequately prepare you for your direct examination by your attorney/client.

CROSS-EXAMINATION

In my experience, most opposing attorneys are civilized. However, some of them can be nasty and abrupt in their cross-examinations. Consider questions like the following, and craft answers that suit your personality and temperament. But listen carefully because a slightly differently worded question may require an entirely different reply. After every question, pause for a beat to give yourself time to form the best answer and to allow your attorney/client time to object. If he or she does object, immediately stop talking until the judge rules. In any case, try not to lose your cool.

Question: How much are you being paid for your testimony?
Answer: *My company charges $xxx per hour for my professional services; my testimony is not for sale.*
Question: Have you ever lied?
Answer: *Of course, but never under oath.*
Question: Have you ever been wrong?
Answer: *Yes, but never on the witness stand.*
Question: Did your attorney/client tell you what to say today?
Answer: *No, of course not. Except to speak slowly and clearly.*

Be especially cautious of compound questions, run-on multiple questions, or ones with multiple parts. Ask to have them rephrased, one question at a time. Politely keep asking for clarifications until you understand the question.

CASES

Accept only cases in your field of expertise, but don't necessarily limit them to your area of specialty. My cases as an electrical engineer included electrocutions, fires of suspected electrical origin, standard of care determinations, equipment failures, arc fault accidents, conveyor accidents, lightning strikes, and others. Yours will be different but also interesting and challenging. I wish you much success in the exciting field of forensic engineering.

CASE STUDY

The following is a narrative synopses of one of my cases. This is not intended to be a formal presentation of a legal action. No confidential or privileged information is revealed. This is simply my recollections, intended only to illustrate an expert's typical involvement in judicial matters.

I was the chief investigator and expert witness on all of the forensic cases of Gaskell Associates, Ltd., and later those of the Gaskell Associates division of Thielsch Engineering until my retirement. However, it was my practice to meet on each case with my senior staff to "brainstorm" the case. This often opened up avenues of investigation that had not previously occurred to me. (I attribute much of my success to hiring others who are smarter than I am.)

Case Study – Standard of Care

My largest forensic case involved the Denver International Airport. This was one of the largest construction projects in the United States for three years in a row and employed more than 10,000 workmen simultaneously. (I was not involved with the famous "Baggage Handling System" problems that delayed the opening of the airport for almost a year.)

Because of what seemed to be an inordinate number of errors and omissions by the main terminal architects and engineers, the city and county of Denver decided to sue. Architects and engineers are not expected to be perfect and are not required to compensate owners for their mistakes, if these professionals meet the standard of care—"that level or quality of service ordinarily provided by other normally competent practitioners of good standing in that field, contemporaneously providing similar services in the same locality and under the same circumstances."

I made six trips to Denver and reviewed files of 143 alleged electrical errors and omissions. I was deposed (oral questioning by the opposing attorney) for six hours and appeared on the witness stand for four hours. Unfortunately, there is no formula to apply regarding standard of care. I needed to evaluate the facts and circumstances and express my professional opinion.

The residents of Denver had endured the presence of rowdy construction workers for ten years and were anxious to get vengeance with a lawsuit. But by the time that the trial actually took place, they had all used the airport, enjoyed the wide concourses, and took pride in the many design awards that the airport received. Their verdict was "not guilty." An unofficial poll later revealed that the jurors' general feeling was "yes, they made a lot of mistakes, but it was a complicated and unique design and anyone would have made slipups."

OUTLINE SUMMARY

- **Forensic engineering:** "The application of the art and science of engineering in matters which are in, or may possibly relate to, the jurisprudence system."
- **Pros:** Interesting work, low liability, high hourly rates, little competition, and advanced payment via retainer.
- **Cons:** Can be stressful and sometimes requires travel.
- **Requirements:** Speaking/teaching ability, writing skills, willingness to prepare, and being a thorough reader.
- **Two types of witnesses:** A fact witness can only tell what they observed, not what others have said (hearsay) and not what they think concerning the case. On the other hand, an expert witness is a technical witness and can express opinions on matters within their expertise.
- **Expert witnesses:** Must be supportive of their opinions and not advocates for their attorney/client.
- **CV:** This is your expert witness resume and includes your present and previous company affiliations, positions and titles, education, certifications, PE licenses, memberships, awards, lectures, articles and papers published, and past forensic cases.
- **Marketing:** Send your CV to the litigation attorneys in your area with a cover letter introducing yourself, join the National Academy of Forensic Engineers (NAFE), contact the companies who frequently hire experts in your field, list in expert directories, and contact insurance companies.
- **Forensic fee agreements:** Should include an up-front, nonrefundable retainer, hourly rates, and a scope that retains you as a consultant until selected as an expert witness.
- **Investigations:** Include site inspections, interviews, examinations of evidence, testing, and reviewing of testing done by others. If you

suspect that evidence has been tampered with or spoiled, immediately inform your attorney/client.

- **White papers:** Case preparation provides an opportunity to prepare narratives of specialized topics to use on future cases and as the basis of articles authored by you.
- **Your case file:** The first two pages should be your case summary and case time and charges.
- **Reports:** Do not write reports until your attorney/client requests. There are three basic types of reports: verbal, outline, and detailed.
- **Depositions:** Are "oral testimony, under oath, taken by the opposing attorney in advance of trial" and are often required of expert witnesses, especially if the case is expected to go to trial.
- **The trial:** The good news is that the vast majority of cases settle; the bad news is that you need to prepare every case as if it will go to trial. Review what to bring to court with your attorney/client. Usually, bring your CV, your fee schedule, time sheets, bills, case summary, and your report. Preparation is the key to success: write out questions and practice responding; try to appear confident, knowledgeable, and likable; consider exhibits and props; sit up straight and lean slightly forward; respond confidently to each question while looking at the jury; present a self-assured demeanor; and try not to let the opposing attorney get you visibly upset.
- **Cross-examination:** Most opposing attorneys are civilized. However, some of them can be nasty and abrupt in their cross-examinations. Craft answers to questions involving your compensation, honesty, being correct, having read everything, having told everything, the reputation of the opposing witness, deleting information from your file, and your attorney/client having told you what to say.
- **Cases:** Accept only cases in your field of expertise, but don't necessarily limit them to your area of specialty. My cases as an electrical engineer included electrocutions, fires of suspected electrical origin, standard of care, equipment failures, arc fault accidents, conveyor accidents, lightning strikes, and others. Yours will be different but also interesting and challenging.

PART 5

CASHING OUT

CHAPTER 17

SELLING YOUR PRACTICE

INTRODUCTION

Learn the pros & cons of just closing; selling to relatives; selling to employees; forming an ESOP; selling to investors; selling to a local engineering firm; or marketing your practice nationally or internationally. Find out what factors determine what your firm is worth and determine step-by- step how to build your exit strategy.

SELLING OPTIONS

Close the doors

Benefits:

This could be done over a number of years while cutting back on hours. It would allow a gradual transition while concentrating on the more interesting projects and the more enjoyable aspects of the work. There would be little concern about being bored because hours would be somewhat flexible. The final tasks would be "construction phase" services, which are usually less demanding. If money was not a factor, this might be the nicest transition choice.

Drawbacks:

Except for the eventual sale of the fixed assets, this option does not provide for a retirement *nest egg*. Also, with the closing of the practice, the company name and reputation ceases, which is a concern for some of us. Although if there were few or no employees, this may be the only viable option. The lasting value of a "service" business is its staff.

Another drawback is the likelihood that your staff will abandon you, because they see that they will soon need a new job.

By the time that I was considering retirement we had grown to a staff of eleven and I had something of value to sell.

Sell to family

Family situations are often awkward and unique. If your children are employed by your business, factors like professional standing and longevity need to be considered. Perhaps you should delay retirement until client loyalties and confidences are transitioned to your children and until they become PEs.

Benefits:

Your name and reputation will endure. This option could have many of the advantages of "closing the doors" and the satisfaction of providing your children an ongoing professional career.

Drawbacks:

It is difficult to charge family members the same price that you would expect an outsider to pay. They often feel entitled and may feel that they should inherit the firm because of their roll in its success. Siblings that are not part of the business will expect their fair share of the value of the firm. Each situation needs to be evaluated on its own merits, but there is a strong likelihood that someone will feel cheated.

Sell to employees

If you have employees who are PEs, they may be candidates to purchase your business.

Benefits:

Your existing employees and your clients already know each other and are used to working together. The existing relationship usually helps to make this kind of transition smooth and successful.

Drawbacks:

As with family, employees often feel entitled because of their contribution to the success of your firm and think that the price should be discounted. Also, they often lack the capital and/or credit needed. It might be necessary for you to accept a small down payment, followed by monthly payments funded by future profits from the business. These payments depend upon the success of the new ownership and may not be secure, if the company fails.

Form an ESOP

ESOPs, "Employee Stock Ownership Plans", are often formed to give employees part ownership in the business over time for free; as part of their overall compensation package. The benefits are improved employee relations, stability and often lower borrowing costs. A disadvantage is that, as more and more of the business becomes employee owned, less of the stock is available to the original equity holders to sell.

However, an ESOP can be formed as a method of selling your business to your employees. Simultaneously, you can also sell or gift a portion of your ownership to any other parties.

Benefits:

- Sale by ESOP is a "stock sale" and often provides substantial tax advantages to the seller.
- After the sale the business can expense retirement plan contributions.
- If you choose to, you can retain operating control, even after the sale.

Drawbacks:

- Full cash payment, at closing, is rare.
- Lenders often require sellers to personally guarantee the loan for a period of time.
- Your eventual full-payment is dependent upon the success of a company that you no longer manage.

If you pursue this method of selling your engineering practice be sure to seek competent and experienced legal and business advice. This method usually works best for firms over 50 employees.

Sell to investors

Selling to investors is unusual, but is an option, especially for larger firms.

Benefits:

- It increases your pool of possible buyers.
- More likely to be an all-cash offer.
- Could be done with an ESOP sale.

Drawbacks:

- Your staff may not be willing to work for non-professionals.
- Future hiring may be impeded for the same reason.
- Outsiders might miss-manage professional services.
- Outsiders might overlook the ethical requirements.
- This arrangement might violate Registration Board requirements.
- It is likely you would be required to stay on through the transition.
- PEs' personal liability might be an issue.

Combine with another local engineering firm

This is sometimes done by small firms as a way of easing into retirement. But, the value that you offer is the clients that you would bring, your reputation, and fixed assets. These would not likely be substantial.

Sell to a local engineering firm

Benefits:
You and your reputation are more of a known entity to a local firm. They can more easily check-you-out, because you both know many of the same people. It expands the buyer's customer base and technical talent. It may expand their services and/or disciplines. It usually simplifies and consolidates management and services.

Drawbacks:
They may attribute less value to your company because you are not the unknown guys from more than 100 miles away.

Also, by marketing only locally, you may miss out on the extra value that a regional or nationwide company might pay to have a branch office in your area, with an existing pool of satisfied clients.

Market the firm nationally

If the first several options mentioned above do not restrict you, I recommend that you market your firm both nationally and locally to both engineering firms and investors.

WHAT IS YOUR COMPANY WORTH

The valuation of an engineering practice is a difficult and complex process; and it is not an exact science. Most engineering firms have few fixed assets. Basically, a firm's value is a function of the estimated "income stream" that it will generate in future years. Of course, no one knows the future, but there are some factors that are good indicators:

- How did your practice do financially over the past three years?
- We often hear that companies sell for a multiple of earnings. But, what are earnings: gross profit; operating profit; net profit; or after tax profit?
- The firm's biggest asset is the number and competence of their technical staff. Are they happy and long term employees? Will they stay under the new ownership?
- Will your departure have a serious negative effect on moral, management, competence, and client relations?
- Has your firm developed specialized services with high profits that are likely to continue after you leave?
- Have you positioned the firm with unique skills or specialized services to give them an advantage for selection for future works?
- What is the overall standing of your firm? Have you developed a reputation for competence and quick & reliable service? Is the firm widely known; locally, statewide, regionally, nationwide, or internationally?
- Are you willing to stay on as an employee to help loyalties transition to the new owners?

YOUR TRANSITION PLAN

Your plan for the sale of your engineering practice will depend upon your circumstances, plans and in part, the status and expectations of the economy at the time.

Here are my step-by-step recommendations:

- Review the selling options listed above and determine what best suites your situation and desires.
- Do some research regarding mergers, acquisitions and the selling of a professional practice. With the advent of the internet, this is now a lot easier to do.
- Contact several Mergers & Acquisition firms and see what services they provide and what fees they charge, including options. After they give you an overall quote, request that they break it down into steps. Try not to commit to the whole package; instead do it one step at a time.
- If you start early enough, an M&A firm may be able to analyze your practice and provide a written plan to position your company for sale in the future.
- In your last few years, contribute the maximum to your retirement plan to build it up while you can. In doing so, you will be contributing generously to your employees' retirement plan, but the alternative is paying more taxes.
- Get a minimum of 2 valuations and carefully review the qualifications of both providers, beforehand.
- In the absence of special restrictions, like family, both a local and nationwide search is probably advisable. The way that we are moving toward a global economy, an international search might not be out of the question.
- Consider specialized mailing lists and e-mail lists.
- The internet has web sites and message boards where you can post for sale information; some of them appear to be free.
- Confidentiality is very important. Don't give out confidential information without an agreement provided by your M&A advisor or your attorney. Don't tell your staff, clients, or anyone else that you are thinking of selling. You don't want your staff or clients to abandon you or your competition to pirate them away.

With perseverance and a little luck you will sell your consulting engineering firm for a fair, but maximum price and retire comfortably and enjoy your grandchildren, hobbies and travel. Perhaps you will even write a book.

Enjoy retirement, you've earned it.

OUTLINE SUMMARY

- **Retirement:** Deciding when to retire is a difficult decision for most consulting engineers because the work is interesting and enjoyable. But do it while your reputation is still intact.
- **Options**:
  - **Close:**
    Pros: Gradual with flexible hours.
    Cons: Little final payout, staff may abandon you, and your company name ceases.
  - **Sell to family**:
    Pros: The pride of providing family a good living.
    Cons: Likely less payout and conflicts with siblings.
  - **Sell to employees:**
    Pros: Existing relationships smooth the way to success.
    Cons: Possible lack of cash/credit, and likely installment sale.
  - **Sell by ESOP:**
    Pros: Tax advantages, and leaves an employee retirement plan.
    Cons: Likely installment payments and personal guarantees.
  - **Sell to investors:**
    Pros: Increases buyer pool, likely all cash, and ESOP possible.
    Cons: Morale, ethics, PE Boards restrictions, your possible continued involvement, and liability.
  - **Sell to a local engineering firm:**
    Pros: Comfortable, and provides clients, staff, and specialties.
    Cons: Possibly lower sales price due to familiarity.
  - **Market the firm nationally:**
    Pros: Maximum sales price and local sale is still possible.
    Cons: Not an option for family or employee sales.

- **Price:** The value of your firm is based on the estimated "income stream" that it will generate in future years. This is indicated by past performance, specialized services, technical staff, special advantages, outstanding skills and reputation, speed and reliability of service, and your willingness to temporarily provide management/advice.
- **Your plan:** Your plan for the sale of your engineering practice will depend upon your circumstances, plans, and the status and expectations of the economy at the time.
- **Step-by-step recommendations:** Review the selling options; do some research; contact several mergers and acquisition firms and see what services they provide and what fees they charge, including options; start early and position your firm for a future sale; contribute the maximum to your retirement plan with company contributions; get a minimum of 2 valuations; do both a local and national search; consider specialized mailing lists and e-mail lists; and post on the internet. Confidentiality is very important.
- **Enjoy retirement:** You've earned it.

FORMS & ILLUSTRATIONS

AFTERWORD

One year after you have read this book and implemented some of my advice, let me know how you are doing. Are you happy in the field of consulting engineering? Are you successful in your own practice? Which of my advice worked best for you? Which of my counsel failed? What new items should I include in my next addition? If you found my guidance valuable, write a review and post it on Amazon.com and BarnesandNoble.com. Also, send the review to me so that I may "excerpt" it in my next addition or in my book promotion. If you didn't like the book, please tell me why. In any case, I wish you much continued success. Keep in touch.

INDEX

V

W

NOTES

NOTES

NOTES

NOTES

www.ingramcontent.com/pod-product-compliance
Ingram Content Group UK Ltd.
Pitfield, Milton Keynes, MK11 3LW, UK
UKHW050944290926
15470UKWH00005B/1636